DECARBONIZING THE WATER SECTOR IN ASIA AND THE PACIFIC

BEST PRACTICES, CHALLENGES, AND OPPORTUNITIES FOR PRACTITIONERS

NOVEMBER 2023

ASIAN DEVELOPMENT BANK

ADB

© 2023 Asian Development Bank
6 ADB Avenue, Mandaluyong City, 1550 Metro Manila, Philippines
Tel +63 2 8632 4444; Fax +63 2 8636 2444
www.adb.org

Some rights reserved. Published in 2023.

ISBN 978-92-9270-452-0 (print); 978-92-9270-453-7 (electronic); 978-92-9270-454-4 (e-book)
Publication Stock No. TIM230531-2
DOI: http://dx.doi.org/10.22617/TIM230531-2

The views expressed in this publication are those of the authors and do not necessarily reflect the views and policies of the Asian Development Bank (ADB) or its Board of Governors or the governments they represent.

ADB does not guarantee the accuracy of the data included in this publication and accepts no responsibility for any consequence of their use. The mention of specific companies or products of manufacturers does not imply that they are endorsed or recommended by ADB in preference to others of a similar nature that are not mentioned.

By making any designation of or reference to a particular territory or geographic area, or by using the term "country" in this publication, ADB does not intend to make any judgments as to the legal or other status of any territory or area.

Please contact pubsmarketing@adb.org if you have questions or comments with respect to content, or if you wish to obtain copyright permission for your intended use that does not fall within these terms, or for permission to use the ADB logo.

Corrigenda to ADB publications may be found at http://www.adb.org/publications/corrigenda.

Notes:
In this publication, "$" refers to United States dollars.
ADB recognizes "China" as the People's Republic of China and "United States of America" as the United States.

On the cover: **Decarbonizing the water sector.** A coherent approach to mainstreaming resilience and decarbonizing water sector operations is vital to ensure effective support and enable policies and investments leading to a net-zero transition, while also supporting development and resilience goals (photos by Abir Abdullah/ADB, Grigoriy Aisenshtat/ADB, Deng Jia/ADB, Lester Ledesma/ADB, Eric Sales/ADB, and Tran Viet Tuan/ADB).

CONTENTS

FIGURES AND BOXES

FOREWORD

The development for a prosperous, inclusive, resilient, and sustainable Asia and the Pacific will be dependent on our capacity to implement measures to adapt to climate change and reduce carbon emissions. From 1970 to 2020, natural hazards in Asia and the Pacific have affected 6.9 billion people and killed more than 2 million. Many of these hazards resulted in water-related disasters such as floods and droughts.[1] Similarly, the water sector—hereby also defined as the "water nexus" to highlight the interconnection and integration with other sectors—is an important entry point globally, as well as at the Asian Development Bank (ADB), to scale up commitments to climate finance and to reduce the impacts of climate change on the most vulnerable countries and individuals.

Water is the primary medium through which the impacts of climate change are felt. Yet, the water sector itself has been widely overlooked as a significant source of greenhouse gas emissions and, as a result, caused missed opportunities to mitigate climate change and support the net-zero carbon transition. The Paris Agreement, established in 2015 at the 21st Conference of the Parties to the United Nations Framework Convention on Climate Change (COP21), does not make a direct reference to water. However, water is identified as the number one priority for many adaptation actions laid out in the nationally determined contributions and is directly or indirectly related to all other priority areas. With the continuing demands for infrastructure investment and the shift to a low-carbon growth trajectory, developing Asia and the Pacific needs to increase knowledge and policy capacity to recognize the economic, social, and environmental costs and opportunities they would face during the net-zero transition.

This ADB guidance note is second in a set of two guidance notes, which will help mainstream resilience, mitigation approaches, and activities in the water sector of ADB's developing member countries (DMCs). The first guidance note, *Mainstreaming Water Resilience in Asia and the Pacific*,[2] was published in August 2022.

This set of two guidance notes supports ADB's wider commitment to scaling up climate actions in the region, including the climate finance target, in line with building climate and disaster resilience—one of the seven operational priorities of ADB's Strategy 2030. In 2021, ADB elevated its ambition to deliver climate financing to its DMCs to $100 billion during 2019–2030, including $66 billion on climate mitigation and $34 billion on climate adaptation. To help achieve these objectives, ADB in 2022 launched at the COP27 the Asia and the Pacific Water Resilience Initiative, commonly known as "RUWR: aRe yoU Water Resilient?," a multipronged endeavor to support local-level authorities of its DMCs to mainstream water security and resilience. Two of the key actions under the Asia and the Pacific Water Resilience Initiative are additional grant and technical assistance mobilization, and rapid capacity building and sharing of knowledge, tools, and solutions toward resilience and decarbonization.

---

[1] United Nations Economic and Social Commission for Asia and the Pacific. 2021. *Resilience in a Riskier World: Managing Systemic Risks from Biological and Other Natural Hazards—Asia-Pacific Disaster Report 2021.* Bangkok.

[2] ADB. 2022. *Mainstreaming Water Resilience in Asia and the Pacific: Guidance Note.* Manila.

ADB remains committed to being a trusted knowledge partner of stakeholders in its DMCs, including governments, development partners, and the private sector—and providing them with the tools, capacities, resources, and information needed to enable transformational change toward a water-secure, green, resilient, inclusive, and sustainable Asia and the Pacific. Alongside the first guidance note, this second guidance note, *Decarbonizing the Water Sector in Asia and the Pacific*, maps challenges and opportunities, and entry points for ADB staff and DMC water sector stakeholders, water nexus entities, and project implementing agencies in the region toward the achievement of resilient and decarbonization outcomes.

Fatima Yasmin
Vice-President (Sectors and Themes)
Asian Development Bank

ACKNOWLEDGMENTS

The guidance note *Decarbonizing the Water Sector in Asia and the Pacific* was prepared by the Strategy and Partnerships Team of the Asian Development Bank (ADB) following extensive consultations with internal and external stakeholders. Alessio Giardino, senior water specialist (Climate Change) and Geoffrey Wilson, senior water resources specialist, led its preparation under the guidance and direction of Neeta Pokhrel, director, Pacific and Southeast Asia Team (former chief of the Water Sector Group) and Satoshi Ishii, director, Strategy and Partnerships Team.

Colleagues from the Climate Change and Sustainable Development Department, Sectors Group, and Private Sector Operations Department provided valuable feedback and support. The team is especially grateful to David Morgado (senior energy specialist), Pedro Miguel Pauleta De Almeida (senior urban development specialist), Hisaka Kimura (advisor, Private Sector Operations Department), Noel Peters (principal investment specialist, Climate Finance), and Mischa Lentz (senior urban development specialist, Finance and Investment) for their inputs.

Craig Hart (executive director of the Pace Energy and Climate Center) served as primary co-author of this report. Casper van der Tak (ADB consultant) peer reviewed the document. Jason Beerman (ADB consultant) edited the document with support from Joanna Brewster (ADB consultant). Gino Pascua provided graphic and visual support to the publication. The team is also grateful to Guy Howard (professor at the University of Bristol) for his feedback and additions.

ABBREVIATIONS

ADB	Asian Development Bank
ASEAN	Association of Southeast Asian Nations
BOD	biochemical oxygen demand
CDM	Clean Development Mechanism
CH_4	methane
CHP	combined heat and power
CO_2	carbon dioxide
COVID-19	coronavirus disease
CPS	country partnership strategy
DMC	developing member country
ECAM	Energy Performance and Carbon Emissions Assessment and Monitoring Tool
GHG	greenhouse gas
G-Res	GHG Reservoir
GWI	Global Water Intelligence
IPCC	Intergovernmental Panel on Climate Change
N_2O	nitrous oxide
NDC	nationally determined contribution
OECD	Organisation for Economic Co-operation and Development
PRC	People's Republic of China
SRI	system of rice intensification
STEEP	Screening Tool for Energy Evaluation of Projects
UNFCCC	United Nations Framework Convention on Climate Change
WASH	water, sanitation, and hygiene
WWTP	wastewater treatment plant

EXECUTIVE SUMMARY

Water is essential for life and is intertwined with virtually every aspect of Asia and the Pacific's development path. Yet, the water sector—hereby also defined as the "water nexus" to highlight the interconnection and integration with other sectors—has been widely ignored as a significant source of greenhouse gas (GHG) emissions. As a result, water has often been overlooked as an opportunity to mitigate climate change.

The Asia and Pacific region faces extraordinary challenges in the water sector both in terms of climate resilience and emissions. One such challenge is that it is home to 60% of the world's population, yet only possesses about a third of global freshwater resources.[1] The Asia and Pacific region accounts for half of global warming forcings in the water infrastructure space (encompassing water supply, wastewater treatment, sanitation, and hygiene) on a 100-year basis, taking into account energy emissions as well as carbon dioxide (CO_2), methane (CH_4), and nitrous oxide (N_2O) emissions from water treatment, wastewater, and sludge treatment for sewer-connected systems and on-site sanitation.[2] Estimates of GHG emissions from wastewater alone indicate that these emissions could be as high as 1.8% of anthropogenic global GHG emissions, roughly equivalent to aviation sector emissions.[3] The broader water sector's emissions may be as much as two to three times higher.[4]

These complex and intertwined challenges of water security, resilience, and decarbonization in Asia and the Pacific require immediate actions that must be complementary, synergizing opportunities not only to pursue adaptation and mitigation goals together, but also to optimize outcomes at the nexus of water use, conservation, land use practices, and overall development goals of the developing member countries (DMCs) of the Asian Development Bank (ADB). All this must be done while finding appropriate cost recovery mechanisms for the DMCs or even reducing costs. Improvements in delivering water, sanitation, and hygiene services to all (and especially under-resourced populations); managing water resources; and adapting all water and sanitation infrastructure and communities they serve for greater resilience are essential to Asia and the Pacific's continued growth and well-being. A coherent approach to mainstreaming resilience and decarbonizing water sector operations is needed to ensure that ADB can effectively support and enable policies and investments in its DMCs, leading to a net-zero transition, while also supporting the DMCs' development and resilience goals. ADB has therefore developed a set of two guidance notes to help mainstream resilience and decarbonization in the water sector in Asia and the Pacific.[5]

---

[1]   United Nations Department of Economic and Social Affairs. International Decade for Action "Water for Life" 2005–2015.

[2]   Global Water Intelligence (GWI). 2022. Mapping Water's Carbon Footprint: Our net zero future hinges on wastewater. Oxford. p. 5.

[3]   GWI. 2022. *Mapping Water's Carbon Footprint: Our net zero future hinges on wastewater*. Oxford. p. 2.

[4]   For example, wastewater treatment plant emissions are estimated as high as 5%, according to World Economic Forum. 2022. How tackling wastewater can help corporations achieve climate goals. 19 October. The wide range in estimates reflects the likely problems with reporting. Country greenhouse gas inventory reports to the United Nations Framework Convention on Climate Change, for example, suggest a lower range, but these numbers do not necessarily focus on the water sector.

[5]   The first of the two guidance notes is ADB. 2022. Mainstreaming Water Resilience in Asia and the Pacific: Guidance Note. Manila.

This guidance note *Decarbonizing the Water Sector in Asia and the Pacific* explores the best practices, challenges, and opportunities that have broad application for supporting climate mitigation and (in parallel) fostering water and sanitation security and resilience in the region. It is primarily designed for water professionals and policymakers supporting or working in ADB's DMCs.

The guidance note evaluates the following five water-related subsectors where ADB is actively supporting its DMCs:

(i)   Water supply
(ii) Sanitation, including wastewater, non-sewered sanitation and drainage, and hygiene
(iii) Water resources management: energy and water storage
(iv) Irrigated agriculture
(v)   Land use and forestry resource management

As regional water infrastructure needs are growing rapidly, and these are long-term assets with several decades of expected service, decarbonizing water infrastructure is essential for avoiding carbon lock-in over the lifetime of the infrastructure. Monetizing those decarbonization opportunities wherever possible will be essential to ensure the provision of affordable water and sanitation services and resources.

Opportunities include increasing water efficiency and promoting water conservation by residential, business and industrial, and agricultural users, which would greatly reduce the volumes of water required to be treated by water supply and wastewater infrastructure, and thereby reduce GHG emissions. Water efficiency reduces water consumption as a measure of the amount of water required for a particular purpose. Water efficiency differs from water conservation in that water efficiency focuses on reducing water waste, not forgoing use.

Water efficiency and conservation efforts must be joined by all sectors of the economy—households, industry, and agriculture—to significantly reduce water sector energy consumption and GHG emissions. Agriculture, which accounts for an estimated 70% of global water consumption,[6] must be a focus of water conservation programs. As the dominant consumer of water, the agriculture sector presents opportunities for high water savings with attendant reductions in energy consumption and GHG emissions.

Technology will play a critical role. Water supply, wastewater treatment, and reservoir management all present opportunities to transition to renewable energy, as well as to adopt innovative GHG reduction or avoidance techniques, such as avoiding anaerobic reactions that produce CH_4, and, if these emissions are unavoidable, to capture the CH_4 and utilize it to produce energy.

For effective and integrated water resources management, planning and design are essential considerations to reduce reservoir GHG emissions. Siting, design, and land use practices should reduce the flooded area used for water storage and cultivation and aim to prevent or reduce the amount of fertilizer and other organic matter that enters reservoirs and that ultimately lead to GHG emissions. These methods include locating siting reservoirs in upland areas with lower temperature microclimates that reduce aquatic plant growth,[7] and promoting agriculture practices near reservoirs such as

---

[6]   Organisation for Economic Co-operation and Development. Water and agriculture.
[7]   R. M. Almeida et al. 2019. Reducing greenhouse gas emissions of Amazon hydropower with strategic dam planning. *Nature Communications.* 10 (1).

Water supply facility in the Marshall Islands. The Ebeye Water Supply and Sanitation Project is linking all households in Ebeye, Marshall Islands to upgraded freshwater and sewage facilities that reduce water leaks and sewage overflows (photo by Eric Sales).

conservation tillage, crop rotation, and cover cropping that can reduce soil erosion, reduce fertilizer use, and avoid fertilizer runoff into reservoirs, as well as sequester carbon in soils.[8]

For Asia and the Pacific, changing large and smallholder farming techniques and building capacity of institutions and farmers to reduce water consumption, and hence emissions, is especially critical for rice production. Rice farming followed by soy farming are among the most GHG-intensive crops because of their high CO_2, CH_4, and N_2O emissions. Possible solutions include substituting these crops with high-value crops and developing new rice variants, adopting water conservation practices for rice cultivation.

Similarly, integrated planning of urban development and agriculture and other land uses with water conservation and decarbonization as twin goals can drive efforts to reduce land conversion in order to protect wetlands and other carbon-storing lands like peatlands, arctic permafrost, salt marshes, mangroves, and eelgrass beds.

While decarbonizing water and sanitation services, significant opportunities also exist to involve the private sector and increase private financing in the water sector, a much-needed and missing piece of the puzzle in the journey toward water security in many DMCs.

The Paris Agreement provides a renewed opportunity to elevate the importance of the water sector in mitigating GHG emissions, as well as synergizing mitigation efforts with adaptation measures. Because

---

[8]   X. Du et al. 2022. Conservation management decreases surface runoff and soil erosion. International Soil and Water Conservation Research. 10 (2).

water resource systems often do not respect national boundaries, transboundary cooperation will be essential to manage water resources based on their geography rather than in an uncoordinated, or even conflicting, piecemeal fashion. Institutions, including ADB, can play an important role in promoting regional approaches to water resources management and convening regional stakeholders, with a view to utilizing these resources to maximize water security, economic development, and decarbonization outcomes. The Paris Agreement's Article 6 cooperation mechanisms for mitigation outcomes offer opportunities to finance water sector mitigation measures through internationally transferred mitigation outcomes, a new type of carbon credit.

Well-founded actions under six mutually reinforcing pillars are proposed, which can set the basis for ADB to further enhance mitigation actions in the water sector through collective actions with partners and DMCs. These pillars are complementary to those presented in the guidance note *Mainstreaming Water Resilience in Asia and the Pacific*. The proposed approach is integrated and inclusive and focuses on long-term climate mitigation outcomes through scaling up financing and building capacity, starting with early engagement of DMCs. The approach provides guidance to the water sector in general, and to ADB in particular, in developing a portfolio of integrated water or cross-sector projects and programs with climate adaptation and mitigation outcomes.

The decarbonization pillars would encourage ADB to

(i)   accelerate early engagement and build demand for water nexus investments leading to mitigation outcomes;
(ii) adopt a water community approach to DMC decarbonization capacity;
(iii) strengthen ADB staff capacity;
(iv) foster knowledge, innovation, and partnerships, primarily through leveraging ADB's strategic position as a center for thought and practice leadership on water resilience and climate mitigation within Asia and the Pacific;
(v)   mobilize finance for water sector decarbonization; and
(vi) spearhead digitalization for decarbonization in the water sector.

The guidance note is organized in four main parts, bookended by an introduction and a conclusion. Chapter 2 reviews water sector decarbonization challenges in Asia and the Pacific and the central role water plays in the region's development. Chapter 3 examines water resilience and decarbonization opportunities and measures. Chapter 4 looks at policies that countries in Asia and the Pacific can adopt to support water resilience and decarbonization. Chapter 5 outlines actions for ADB under the six pillars to further enhance mitigation actions in the water sector.

Beneficiaries of Tonle Sap Rural Water Supply and Sanitation Project in Kampong Chhnang Province, Cambodia. The project promotes the use of safe water and hygiene, stops open defecation, constructs improved latrines, and maintains the water sources (photo by Eric Sales).

1 Introduction

Water has always been recognized as essential. Yet, the water sector itself has been widely excluded as a significant source of greenhouse gas (GHG) emissions. As a result, the water sector has often been overlooked as an opportunity to mitigate climate change.

The Paris Agreement provides a renewed opportunity to elevate the importance of the water sector in mitigating GHG emissions, as well as synergizing mitigation efforts with adaptation measures. The Paris Agreement supports countries in initiating their own sector-wide mitigation approaches through nationally determined contributions (NDCs), the foundation of Paris Agreement efforts to decarbonize and adapt to climate change. The Paris Agreement's Article 6 cooperation mechanisms for mitigation outcomes are intended to help financially support the implementation of NDCs. Yet, NDCs largely ignore water as a factor in decarbonization planning.[1]

Even though the Intergovernmental Panel on Climate Change (IPCC) has recognized the critical carbon sequestration services provided by wetland ecosystems toward achieving Paris Agreement goals,[2] only 12 countries have adopted NDCs incorporating wetlands[3] as a mitigation measure during the first iteration of NDCs made during 2015–2019.[4] In the current round of NDCs to be concluded by 2025, while water-related activities

---

[1]   T. Rudebeck et al. 2022. Chapter 3: Governance context of water-related climate mitigation measures. Box 3.2. In M. L. Ingemarsson et al., eds. *The Essential Drop to Net-Zero: Unpacking Freshwater's Role in Climate Change Mitigation*. Stockholm: Stockholm International Water Institute.

[2]   IPCC. 2019. *IPCC Special Report on Climate Change, Desertification, Land Degradation, Sustainable Land Management, Food Security, and Greenhouse gas fluxes in Terrestrial Ecosystems (Approved Draft)*. Geneva.

[3]   For the purpose of this guidance note, the term wetlands is used as a generic term for carbon-storing lands and includes peatlands, arctic permafrost, salt marshes, mangroves, and eelgrass beds.

[4]   N. Anisha et al. 2020. *Locking Carbon in Wetlands: Enhancing Climate Action by Including Wetlands in NDCs*. Corvallis, Oregon and Wageningen, Netherlands: Alliance for Global Water Adaptation and Wetlands International.

feature more prominently, water remains primarily confined to adaptation measures, and few enhanced NDCs propose water-sector emission mitigation strategies (footnote 1).

To help elevate the water sector as a priority for combined mitigation and adaptation efforts, this guidance note evaluates the critical role that water plays in decarbonization, identifies opportunities for synergies between mitigation and more resilient water resources, and proposes measures that can be taken to advance the pursuit of these opportunities.

This guidance note is primarily designed as a document for water professionals and policymakers supporting or working in the developing member countries (DMCs) of the Asian Development Bank (ADB). The guidance note describes enablers and best practices that have broad application for fostering climate resilience in the water sector. For a summary of the findings provided in this guidance note we also refer to the Asian Development Outlook 2023 background paper on decarbonizing the water sector.[5]

---

[5]   A. Giardino, G. Wilson, and C. Hart. 2023. Decarbonizing the Water Sector. Background Paper for Asia in the Global Transition to Net Zero. Asian Development Outlook Thematic Report 2023. Manila: ADB.

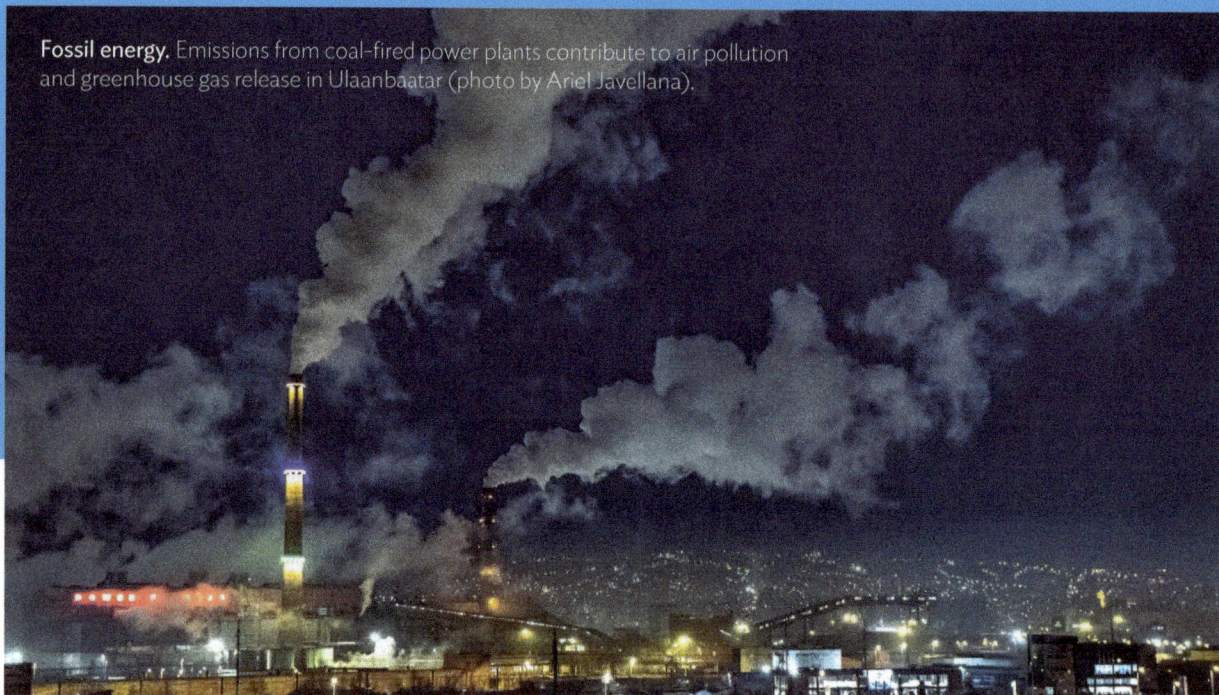

Fossil energy. Emissions from coal-fired power plants contribute to air pollution and greenhouse gas release in Ulaanbaatar (photo by Ariel Javellana).

2 Water Mitigation and Development Challenges in Asia and the Pacific

2.1 Overview of Water Mitigation and Development Challenges

The Asia and Pacific region faces diverse and complex water decarbonization challenges. In the water sector, purchasing carbon-intensive electricity used in an entity's operations and producing GHG emissions from operations—primarily carbon dioxide (CO_2), methane (CH_4), and nitrous oxide (N_2O)—are both important aspects of the decarbonization challenge. CH_4 is produced by water sector operations in high volumes and is an extremely potent GHG over the short term, with a global warming potential of 82.5 times that of CO_2 over a 20-year period, and 29.8 times that of CO_2 over a 100-year period.

The 20-year global warming potential of emissions is extremely relevant because it is what occurs over the next 20 years that, to a large extent, will determine the future of the planet. The global warming potential of N_2O is 273 times that of CO_2 over both 20-year and 100-year periods.

The region's efforts to reduce its GHG emissions are crucial to the entire world's efforts to fight global climate change. The Asia and Pacific region accounts for one-third of global gross domestic product and 42% of global GHG emissions from fossil energy consumption. The People's Republic of China (PRC) and India are among the world's top three GHG-emitting countries, with the PRC alone accounting for more than 27% of global emissions based on fossil fuel combustion.[6]

---

6   World Bank. Open Data (accessed 5 December 2022).

The magnitude of Asia and the Pacific's water decarbonization challenge is further complicated by the region's immense scope and geographical diversity. The region covers a quarter of the world's land area,[7] incorporating two of the world's great oceans as well as the world's highest mountain ranges, which are critical glacial water sources, feeding several of the world's most ecologically and economically significant rivers, wetlands, and river delta systems.

The Asia and Pacific region's water ecosystem magnitude and complexity is only surpassed by the demands on its water resources and the diverse development challenges that the region presents. The region is home to 60% of the world's population, yet only possesses about a third of global freshwater resources.[8] The region is home to the PRC and India—the world's most populous countries—as well as many of its smallest microstates.[9] The Pacific's small island developing states and Asia's coastal states are especially vulnerable to the water-related impacts of climate change, including sea-level rise, salination of freshwater resources, increased risks of flooding, droughts, and food insecurity.

Further complicating the decarbonization challenge is the fact that while the Asia and Pacific region almost halved extreme poverty during 2000–2019, extreme poverty still affects significant portions of the population in the region.[10] Since 2020, the coronavirus disease (COVID-19) pandemic has contributed to widening disparities in living conditions and food security for the most vulnerable in the region,

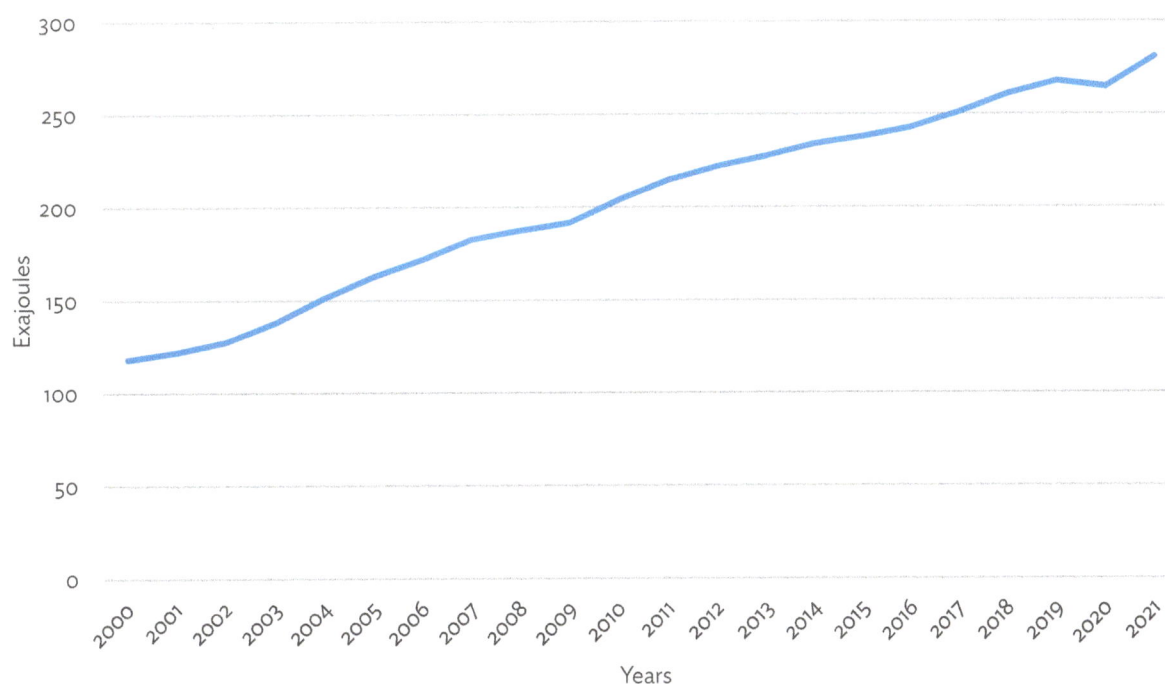

Figure 1: Primary Energy Consumption in Asia and the Pacific, 2000–2021

Source: BP. 2022. *bp Statistical Review of World Energy 2022*. London.

---

7   Food and Agriculture Organization of the United Nations. Overview of the Asia-Pacific Region.
8   United Nations Department of Economic and Social Affairs. International Decade for Action "Water for Life" 2005–2015.
9   World Population Review. Smallest Countries.
10   Organisation for Economic Co-operation and Development (OECD). Society at a Glance: Asia/Pacific 2022—Poverty.

particularly poorer households, rural populations, and women and children.[11] The high cost of energy and economic inequality will only make the region's decarbonization efforts all the more challenging to achieve.

Water is intertwined with virtually every aspect of Asia and the Pacific's development path—industrialization, urbanization, energy consumption, and consumer lifestyles—reshaping the land and water resources in the process.

Water sector emissions are closely linked to energy consumption. Asia and the Pacific remains one of the most rapidly growing regions of the world, with steadily increasing energy consumption and corresponding GHG emissions.

As the region continues to develop, its water sector emissions will continue to increase if efforts to decarbonize are not realized. Figure 1 shows the trend in primary energy consumption for Asia and the Pacific from 2000 to 2021, when the region's energy consumption increased on average by about 7% each year, from almost a third of global consumption to roughly half.[12]

With the increase in primary energy consumption largely because of the growth in the electricity generation and transportation sectors, the region's GHG emissions from energy consumption similarly steadily increased from 2000 to 2021, as shown in Figure 2, which presents CO_2 fossil fuel emissions.

Figure 2: Energy Consumption Carbon Dioxide Emissions in Asia and the Pacific, 2000–2021

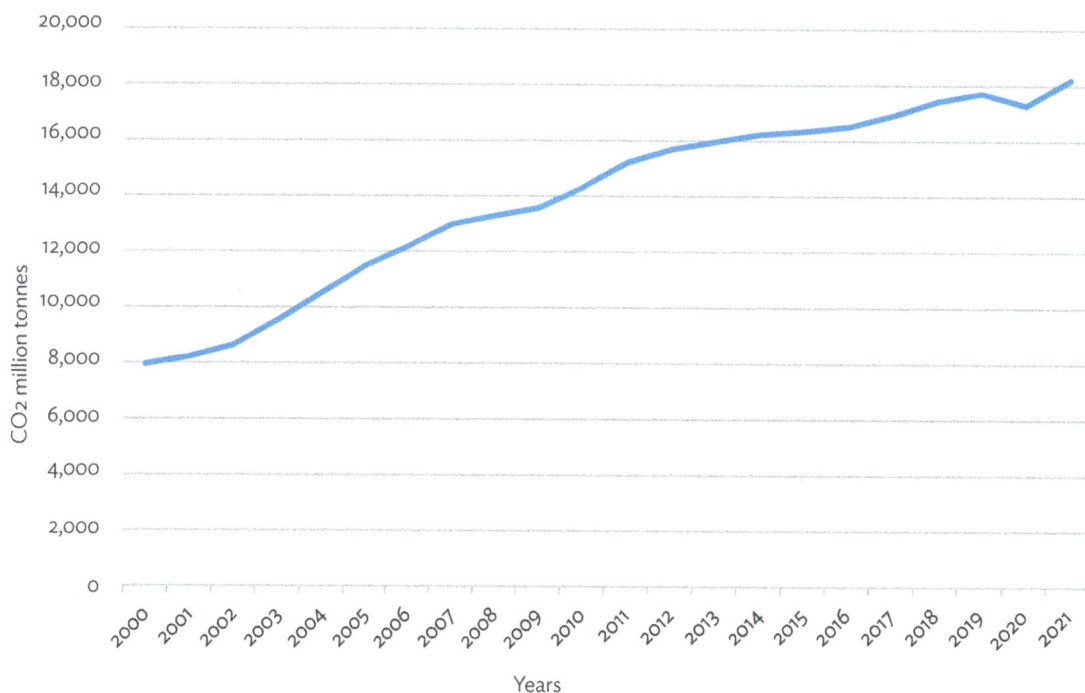

CO_2 = carbon dioxide.
Source: BP. Statistical Review of World Energy — all data, 1965–2021 (accessed 1 June 2023).

---

[11]   United Nations Economic and Social Commission for Asia and the Pacific. 2022. Asia and the Pacific SDG Progress Report 2022: widening disparities amid COVID-19. Bangkok.
[12]   BP. Statistical Review of World Energy.

2.2 Defining the Water Sector

In assessing opportunities to mitigate GHG emissions while enhancing the resilience of the water sector, this guidance note evaluates five subsectors of the water sector, which covers ADB's characterization of the water sector. This includes the water, sanitation, and hygiene (WASH) subsector, addressing water supply separately from wastewater, sanitation, and hygiene. Water supply includes primary water abstraction, treatment, and distribution as well as potentially desalination. Wastewater, sanitation, and hygiene comprise wastewater discharged by households and industry and its conveyance and treatment, as well as sanitation and hygiene practices that pollute water resources. The water sector also includes the management of large water resources in reservoirs and canals and/or channels and abstracting, treating, and transporting irrigation water for agriculture. Finally, it includes land use changes that affect

water bodies, such as wetlands that store significant carbon, and their loss through land use development resulting in significant GHG emissions. For the purposes of this guidance note, hydropower is not considered part of the water sector, as it is commonly covered under ADB's energy sector; however, this guidance note does discuss emissions from reservoirs irrespective of their purpose.

In summary, for this guidance note, the water sector constitutes the following subsectors:

(i)   Water supply
(ii) Wastewater, sanitation, and hygiene
(iii) Water resources management: energy and water storage
(iv) Irrigated agriculture
(v) Land use and forestry resource management

Estimates of GHG emissions from wastewater alone indicate that these emissions could be as high as 1.8% of anthropogenic global GHG

Figure 3: Global Greenhouse Gas Emissions by Sector, 2016

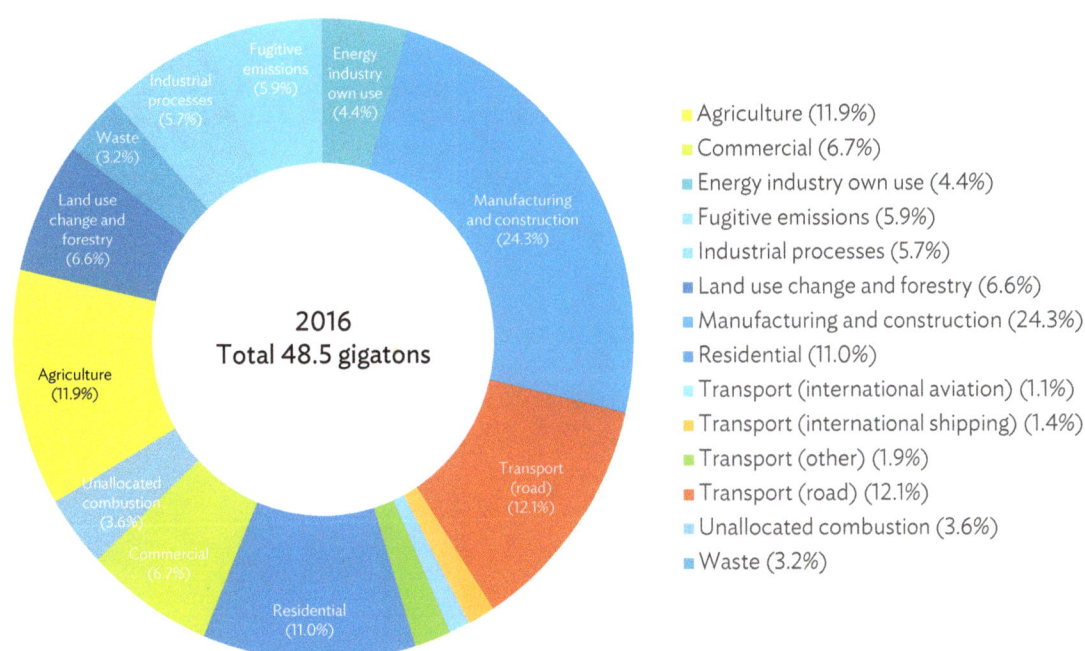

2016
Total 48.5 gigatons

- Agriculture (11.9%)
- Commercial (6.7%)
- Energy industry own use (4.4%)
- Fugitive emissions (5.9%)
- Industrial processes (5.7%)
- Land use change and forestry (6.6%)
- Manufacturing and construction (24.3%)
- Residential (11.0%)
- Transport (international aviation) (1.1%)
- Transport (international shipping) (1.4%)
- Transport (other) (1.9%)
- Transport (road) (12.1%)
- Unallocated combustion (3.6%)
- Waste (3.2%)

Source: Earthcharts.org.

emissions, roughly equivalent to aviation sector emissions.[13] The broader water sector's emissions may be as much as two to three times higher.[14] An overview of GHG emissions by sector is shown in Figure 3. Although the water sector is not specifically mentioned, it is included in several categories, including residential, commercial, waste, land use change and forestry, and agriculture.

Water supply includes producing, abstracting, treating, pumping, and distributing water, primarily for residential, commercial, and industrial use, and mainly in an urban context. These processes consume carbon-intensive electricity.

2.3 Water Supply

Producing, Abstracting, Treating, Pumping, and Distributing Water

Given the high emission intensity of water treatment and desalination operations, any water physically lost by a water utility in the process of providing water to the consumer is not only a financial loss to the utility that does not provide the intended benefit, but also a loss to the environment in the form of GHG emissions through the wasted energy in treating and pumping this lost water. Nonrevenue water loss is pervasive and significant in the Asia and Pacific region. ADB estimates that nonrevenue water loss because of physical losses (including leaks, bursts and overflow); unauthorized uses (including theft and illegal connections, and inaccurate metering); or authorized uses such as for fire suppression, averages 30% of daily suppression and can reach as high as 65% for some utilities.[15]

Aerial shot of the 26-cubic-meter water reservoir in Bharabhid village, Bajhang District, Nepal. The Building Climate Resilience of Watersheds in Mountain Eco-Regions project is improving water management and providing water supply for use in the home and in agriculture in about 100 communities in Nepal (photo by Gerhard Joren).

---

[13]   Global Water Intelligence (GWI). 2022. *Mapping Water's Carbon Footprint: Our net zero future hinges on wastewater.* Oxford. p. 2.
[14]   For example, wastewater treatment plant (WWTP) emissions are estimated as high as 5%, according to World Economic Forum. 2022. *How tackling wastewater can help corporations achieve climate goals.* 19 October.
[15]   ADB. 2007. *Water Brief: Curbing Asia's Nonrevenue Water.* Manila.

Redressing physical (real) water loss—especially leaks and overflow, which are a major component of nonrevenue water—is therefore critical to significantly reducing GHG emissions from the water supply subsector. This will have the added benefit of reducing water utilities' energy expenditure, which is typically a significant component of their total operating costs, and thereby helping to ensure these utilities are financially sustainable to attract capital and invest in further water efficiency and decarbonization improvements. For example, pressure-regulating devices decrease downstream pressure, which will reduce leaks and is likely to save energy and extend the life expectancy of the asset (pipe).

Water Desalination and Water Processing

As water resources become scarcer, water desalination and wastewater processing for potable reuse are increasingly being considered and adopted. Both processes involve higher energy requirements than primary water sources, resulting in associated higher GHG emissions.

As much as 60%–75% of the cost of desalinating seawater is due to energy consumption.[16] Similarly, the energy requirements for treating wastewater to potable standards for reuse are also higher than ordinary water treatment processes because of the additional steps required to ensure water health and safety standards are achieved. [17]

2.4 Wastewater Treatment, Sanitation, and Hygiene

Wastewater includes collecting (on-site and for off-site treatment), pumping, transporting (for off-site facilities and disposal of sludge), and treating wastewater; it also includes ultimate disposal—typically discharge to water bodies and deposit of sludge in landfill. Wastewater treatment produces CO_2 emissions from energy consumption, and the release of CO_2, CH_4, and N_2O from the treatment processes.

Treatment processes generate GHG emissions through aerobic decomposition (generating primarily CO_2), anaerobic decomposition (generating CH_4), and nitrification-denitrification (generating N_2O). The three main wastewater pathways are

(i)   wastewater treated via a wastewater treatment plant (WWTP), requiring energy and emitting GHGs;
(ii) wastewater released to the environment untreated and emitting GHGs, with emissions from untreated wastewater being an estimated three times higher than emissions from conventional WWTPs on a volumetric basis;[18] and
(iii) wastewater accumulated within on-site sanitation and not emptied and emitting GHGs.

Because of the energy required to operate conventional WWTPs, transitioning from on-site sanitation to centralized sanitation typically involves CO_2 emissions from the consumption of electricity, leading possibly to higher overall GHG emissions. These emissions can be mitigated to potentially achieve overall reductions if aggressive mitigation measures are taken.

Wastewater Treatment

In developing Asia and the Pacific, WWTP facilities are increasingly priorities for development. WWTPs are essential for sanitation and hygiene, especially in urbanized areas; can provide water for industry and agricultural irrigation and other non-potable reuse; and are even increasingly seen as a provider of potable water in severe scarcity conditions.[19]

[16]   C. Bhodi. 2022. Global instability begets energy instability begets water instability. *GWI Magazine.* 23 (10). p. 17.

[17]   *GWI Magazine.* 2022. Utility Performance: Colorado utilities welcome new direct potable reuse regulation. 23 (12). pp. 30–32.

[18]   R. Giné-Garriga et al. 2022. Chapter 4: Mitigation measures in drinking water and sanitation services. In M. L. Ingemarsson et al., eds. *The Essential Drop to Net-Zero: Unpacking Freshwater's Role in Climate Change Mitigation.* Stockholm: Stockholm International Water Institute.

[19]   *GWI Magazine.* 2022. Indian cities look to recycling to boost drinking water supply. Utility Performance: Looking for new sources of water in Chennai. 23 (10). pp. 5, 15, and 32–33.

WWTPs are highly energy intensive because of electricity requirements, and thus significant GHG emitters. For example, the national average wastewater treatment electricity consumption rate in the PRC is as high as about 0.38 kilowatt-hours per cubic meter.[20] Electricity consumption typically accounts for about 43% of the emissions of WWTPs.[21] Their high energy consumption also presents major barriers to affordability for some developing countries.

By one estimate, direct GHG emissions at WWTPs accounts for 1.6% of global GHG emissions, with wastewater treatment responsible for 5% of the total global non-CO_2 GHG emissions (e.g., CH_4 and N_2O).[22] Estimates by Global Water Intelligence (GWI) suggest that emissions from wastewater may be as high as 1.8% of global GHG emissions (footnote 13). This estimate includes anthropogenic direct emissions of CH_4 and N_2O from WWTP operations and indirect emissions from their electricity consumption. In this study, the wastewater emissions estimate does not include

- emissions from connected upstream water collection infrastructure and sewers, especially rising mains (force mains);

- emissions for WWTP and related infrastructure construction and materials (steel and concrete);
- emissions from the production of chemicals and other materials used by WWTPs;
- water utility non-water emissions, such as administration and business-related travel;
- CO_2 emissions from fossil sources embedded in sludge released during processing;[23]
- vehicle emissions for sludge and biosolids transportation;[24] and
- emissions from sludge disposal, such as CH_4 from landfills and N_2O in land applications.[25]

Sanitation and Hygiene

Today, 7% of the world's population lacks access to safely managed sanitation services, resorting to open defecation in soil, water bodies, and streets. Although the IPCC assumes open defecation emits no GHG emissions based on the assumption that anaerobic conditions are unlikely to occur,[26] research suggests that open defecation emits about 20% of the CH_4 emissions associated with pit latrines and a small amount of N_2O emissions.[27] Open defecation also spreads disease; pollutes water, soil, and food; and undermines development efforts.[28]

---

[20]   J. Yang and B. Chen. 2021. Energy efficiency evaluation of wastewater treatment plants based on data envelopment analysis. *Journal of Applied Energy*. 289.

[21]   GWI. 2022. Mapping Water's Carbon Footprint: Our net zero future hinges on wastewater. Oxford. pp. 2 and 6.

[22]   L. Lu et al. 2018. Wastewater Treatment for Carbon Capture and Utilization. *Nature Sustainability*. 1 (12). pp. 750–758.

[23]   Y. Law et al. 2013. Fossil organic carbon in wastewater and its fate in treatment plants. *Water Research*. 47 (14). pp. 5270–5281; and A. G. Schneider, A. Townsend-Small, and D. Rosso. 2015. Impact of direct greenhouse gas emissions on the carbon footprint of water reclamation processes employing nitrification-denitrification. *Science of the Total Environment*. 505. pp. 1166–1173.

[24]   H. Yoshida, J. J. Gable, and J. K. Park. 2012. Evaluation of organic waste diversion alternatives for greenhouse gas reduction. *Resources Conservation and Recycling*. 60. pp. 1–9.

[25]   W. Chen et al. 2022. *The GHG mitigation opportunity of sludge management in China*. Environmental Research. 212.

[26]   According to IPCC guidelines, "Open defecation is not considered as a source of CH_4, as anaerobic conditions are considered unlikely." IPCC. 2019. *Refinement to the 2006 IPCC Guidelines for National Greenhouse Gas Inventories*. Geneva. p. 6. 11. CO_2 and N_2O emissions are not reported and assumed insignificant as well.

[27]   K. Shaw, C. Kennedy, and C. C. Dorea. 2021. Non-Sewered Sanitation Systems' Global Greenhouse Gas Emissions: Balancing Sustainable Development Goal Tradeoffs to End Open Defecation. *Sustainability*. 13 (21). In addition, using Kampala in Uganda as an example, uncontained fecal waste and other forms of on- and off-site sanitation are contributing to high levels of GHG emissions. J. Johnson et al. 2022. Whole-system analysis reveals high greenhouse-gas emissions from citywide sanitation in Kampala, Uganda. *Communications Earth and Environment*. 3 (1).

[28]   GWI. 2022. *Mapping Water's Carbon Footprint: Our net zero future hinges on wastewater*. Oxford. p. 16.

Another 50% of the world's population relies on on-site sanitation, almost two-thirds of which are wet and dry pit latrines, and the remainder septic tanks. On-site sanitation systems are prone to leakage and flooding because of poor maintenance and weather events, resulting in escape of waterborne diseases and pollution to soil and water.[29] Beyond their risk to health and contamination of safe drinking water and soils, these systems emit significant GHG emissions (roughly 94% of which comprise CH_4) during containment, with the remainder N_2O emissions and some CO_2 emissions because of energy consumption when they are periodically emptied (pumping- and transport-related emissions).[30]

Overall, based on GWI estimates, on-site sanitation produces 32% of global water infrastructure emissions, and treatment of wastewater and sludge produces another 30% of global water infrastructure emissions; in these cases, water infrastructure is defined as drinking water treatment and distribution, sewage treatment and on-site sanitation, and energy emissions from sewage collection (footnote 21). Therefore, achieving Sustainable Development Goal 6's objective of providing everyone with sanitation infrastructure may significantly increase GHG emissions, unless the WASH subsector is put on a path toward decarbonization.

BOX 1

Sanitation, Greenhouse Gas Emissions, and Public Health

Some greenhouse gas (GHG) emissions from sanitation systems are unavoidable because when organic matter decomposes it produces carbon and nitrogen species. Concerns about GHG emissions from sanitation must be balanced with the public health risks associated with an absence of sanitation. Inadequate sanitation causes over a quarter of a million children aged under 5 years to die from diarrheal disease in the Asia and Pacific region each year. Diseases caused by poor sanitation also lead to GHG emissions as people must buy medicines and use health-care facilities, both of which lead to an increased global carbon footprint. Achieving Sustainable Development Goal targets on sanitation will offer substantial health benefits and should remain the focus of attention. Efforts to reduce sanitation-related GHGs must not compromise public health. Estimates of emissions from on-site sanitation remain highly uncertain as there is limited empirical data from low- and middle-income countries in tropical climates. Studies are improving estimates and these point to relatively high contributions of methane from poorly designed and managed systems. Reducing these emissions will require better management of existing systems by moving sludge rapidly to treatment plants to avoid unmanaged anaerobic conditions forming, and allowing GHG capture and resource recovery.

- Guy Howard, *University of Bristol*

Source: Asian Development Bank.

---

29   GWI. 2022. *Mapping Water's Carbon Footprint: Our net zero future hinges on wastewater.* Oxford. p. 15.
30   GWI. 2022. *Mapping Water's Carbon Footprint: Our net zero future hinges on wastewater.* Oxford. p. 6.

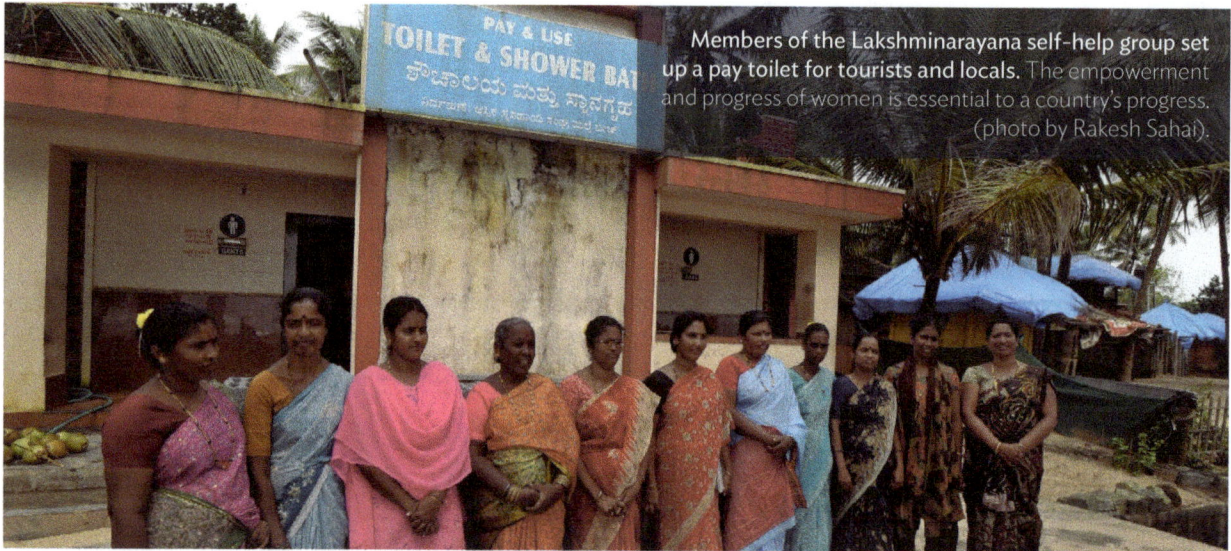

Members of the Lakshminarayana self-help group set up a pay toilet for tourists and locals. The empowerment and progress of women is essential to a country's progress. (photo by Rakesh Sahai).

BOX 2

Water Supply, Wastewater Treatment, Sanitation, and Hygiene Subsectors Emissions

As Asia and the Pacific water infrastructure needs are growing rapidly, and these are long-term assets of several decades of expected service, decarbonizing water infrastructure is essential to avoiding carbon lock-in over the lifetime of the infrastructure. Monetizing those decarbonization opportunities wherever possible will be essential to ensure the provision of affordable water supply, wastewater treatment, sanitation, and hygiene subsector services. Water sector mitigation measures can enhance meeting the United Nations Sustainable Development Goals for public health and hygiene, and resiliency to climate change, and thus represent an investment that synergizes mitigation, adaptation, and sustainability.

Asia and the Pacific accounts for half of global warming potential in the water infrastructure space (encompassing water supply, wastewater, sanitation, and hygiene) on a 100-year basis.[a] Carbon dioxide from energy consumption, primarily because of pumping and running treatment operations, accounts for the greatest share of greenhouse gas emissions in the water infrastructure space weighted based on 100-year global warming potential, followed by methane and nitrous oxide emissions from water treatment, wastewater and sludge treatment for sewer-connected systems, and on-site sanitation (footnote a).

Figure: Water, Sanitation, and Hygiene: 100-Year Global Warming Potential Contribution by Region

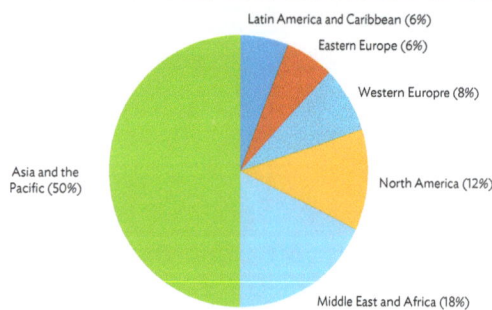

- Latin America and Caribbean (6%)
- Eastern Europe (6%)
- Western Europre (8%)
- North America (12%)
- Middle East and Africa (18%)
- Asia and the Pacific (50%)

Figure: Water, Sanitation, and Hygiene: Global Warming Potential Contribution by Gas

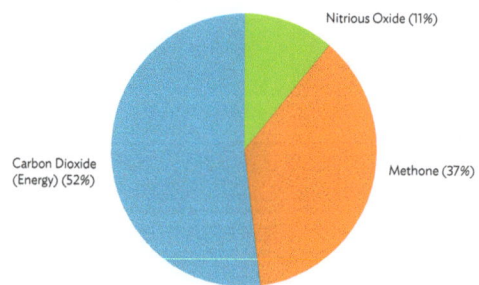

- Nitrious Oxide (11%)
- Methone (37%)
- Carbon Dioxide (Energy) (52%)

Source: Global Water Intelligence. 2022. *Mapping Water's Carbon Footprint: Our net zero future hinges on wastewater*. Oxford.

Source: Global Water Intelligence. 2022. *Mapping Water's Carbon Footprint: Our net zero future hinges on wastewater*. Oxford.

continued on next page

Box 2   *continued*

Energy Consumed by Water Sector Infrastructure

Because of the energy intensity of providing clean water and treating wastewater, electricity emissions heavily influence overall water sector infrastructure emissions.

The provision of clean water for households, agriculture, and industry accounts for about 38% of global water infrastructure emissions, which is entirely due to electricity used for clean water abstraction, treatment, and network operations, as well as desalination processes. The treatment of wastewater and sludge produces another 30% of global water infrastructure emissions, of which about 43% is due to electricity to operate pumps and processing equipment to remove contaminants, and the remainder from chemical processes. The remaining 32% of global water infrastructure emissions are from on-site sanitation, mainly the operation of septic tanks and pit latrines, which are not energy intensive—most of these emissions are from chemical decomposition processes.

Asia and the Pacific's energy sector is the most carbon intensive globally[b] because of its high dependence on coal and other fossil fuels. Coal-fired power accounts for almost half of the region's primary energy consumption. Coal-fired power accounts for two-thirds of energy consumption in the People's Republic of China and three-quarters in India.[c]

The figures below show energy consumption and carbon dioxide emissions due to energy consumption from water infrastructure in the water, sanitation, and hygiene subsector by region. These figures do not include emissions from chemical processes in the water treatment process, nor land use and other non-electricity consumption sources. The Asia and Pacific region is the largest energy-consuming and carbon dioxide-emitting region globally.

Figure: Water, Sanitation, and Hygiene Electricity Consumption by Region

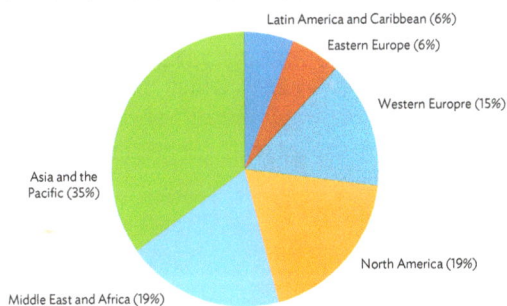

Latin America and Caribbean (6%)
Eastern Europe (6%)
Western Europre (15%)
North America (19%)
Middle East and Africa (19%)
Asia and the Pacific (35%)

Source: Global Water Intelligence. 2022. *Mapping Water's Carbon Footprint: Our net zero future hinges on wastewater.* Oxford.

Figure: Water, Sanitation, and Hygiene Carbon Dioxide Emissions from Electricity Consumption by Region

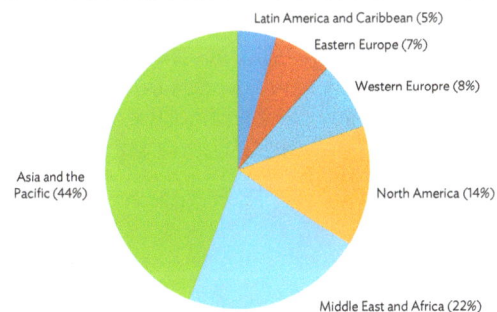

Latin America and Caribbean (5%)
Eastern Europe (7%)
Western Europre (8%)
North America (14%)
Middle East and Africa (22%)
Asia and the Pacific (44%)

Source: Global Water Intelligence. 2022. *Mapping Water's Carbon Footprint: Our net zero future hinges on wastewater.* Oxford.

---

[a] Global Water Intelligence. 2022. *Mapping Water's Carbon Footprint: Our net zero future hinges on wastewater.* Oxford. p. 5.

[b] International Energy Agency. 2023. *Electricity Market Report 2023.* Paris.

[c] BP. Statistical Review of World Energy.

Source: Asian Development Bank.

2.5 Water Resources Management: Energy and Water Storage

Water resources management includes storing water in reservoirs and conveying water in large canal systems. Energy consumption primarily through pumping irrigation water is important. Reservoirs emit CH_4 and CO_2 from the water body and land use changes, especially in the case of large hydropower installations.

Hydropower reservoirs, a significant source of electricity in the region, are an important source of renewable energy, accounting for 14% of the region's electricity generation. However, the reservoirs created by damming rivers can damage the environment and cause the release of CO_2 and CH_4.[31] These emissions are the result of plants and other organic matter in the reservoir's bed material decaying and releasing CO_2 and CH_4 in the process. Additional emissions may come from land clearing and possible burning of the above-ground vegetation to prepare for reservoir filling and from building materials and operations used for dam construction.

Estimating reservoir emissions is subject to large uncertainties because of lack of data and a standardized monitoring protocol. As a result, estimates of life-cycle emissions intensities over different hydropower generation facilities vary

---

BOX 3

Energy and Water in Asia and the Pacific

Reliance on coal and hydropower also makes the energy generation sector reliant on water. Coal mining and washing, and plant operations (including steam production and cooling) all consume water. Hydropower is entirely reliant on water availability for operations and must also balance downstream water user needs. Both the water and energy sectors are vulnerable to water scarcity. With the exception of renewables, which only accounts for 6% of global energy consumption, the entire energy sector is vulnerable to the water impacts of climate change. In turn, Asia and the Pacific's extreme water vulnerabilities puts the region's energy sector at heightened risk.

Figure: Asia and Pacific Primary Energy Consumption by Fuel, 2021

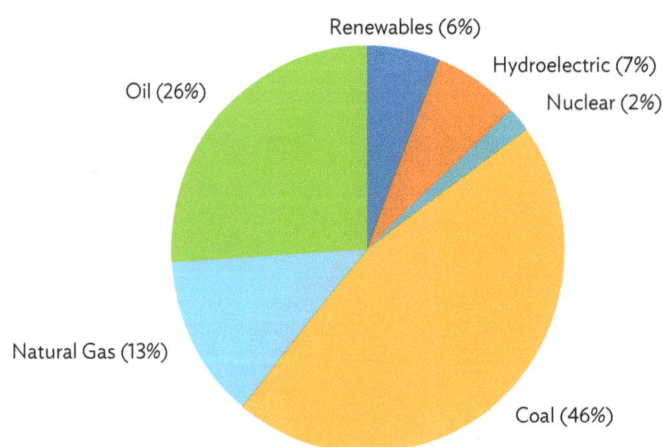

- Renewables (6%)
- Hydroelectric (7%)
- Nuclear (2%)
- Oil (26%)
- Natural Gas (13%)
- Coal (46%)

Source: BP. 2022. *bp Statistical Review of World Energy 2022*. London.

Source: Asian Development Bank.

---

[31]   I. M. Drupady et al. 2021. *Resilience of Renewable Energy in Asia Pacific to the COVID-19 Pandemic*. Singapore: National University of Singapore, Energy Studies Institute.

widely. The IPCC estimated median hydropower GHG emission intensity at 24 grams of CO_2 equivalent per kilowatt-hour, but with a wide range of emission intensities from 0 to 2,220 grams of CO_2 equivalent per kilowatt-hour.[32] For comparison, modern to advanced high-efficiency coal power plants emit 710–950 grams of CO_2 equivalent per kilowatt-hour, and natural gas combined-cycle plants emit 410–650 grams of CO_2 equivalent per kilowatt-hour with high variability in CH_4 emissions from gas production.[33]

2.6 Irrigated Agriculture

Irrigated agriculture includes conveying and distributing water for agricultural production. Energy consumption due to pumping and CH_4 releases from certain crop production are important.

Agriculture accounts for about 12% of global GHG emissions, as shown in Figure 3.

Among crops, rice followed by soy are the among the most GHG intensive. Traditional wet rice farming releases CH_4 as the result of anaerobic decomposition of soil organic matter, resulting in 4.5 kilograms of CO_2 equivalent per kilogram of rice.[34] CH_4 from rice cultivation accounts for about 1.3% of all human-induced GHG emissions globally.[35]

An increasingly significant driver of food-related water consumption and GHG emissions is a pattern of increasing food waste. Increasing food losses are driven by rising household income,

industrialization, and developmental levels.[36] Both developed and industrialized countries suffer from food loss including during the production, storage or transportation, and retailing stages because of agricultural practices; poor infrastructure for storage, processing, and transport; and other value chain constraints. However, food loss is increasingly occurring in industrializing Asia at the consumption stage to levels approaching those in developed countries.[37]

2.7 Land Use and Forestry Resource Management

Land use change, driven by urban development as well as land conversion for agriculture and animal husbandry, causes significant GHG emissions, primarily resulting from wetlands conversion and deforestation. According to one reference, agriculture, forestry, and other land use accounts for 20%–24% of global GHG emissions.[38] This is roughly in the range provided in Figure 3 that indicates a total of 18.5% after adding GHG emissions from the agriculture and land use change and forestry sectors.

Loss of wetlands is an especially important driver of GHG emissions. Wetlands, including peatlands, mangroves, salt marshes, eelgrass beds, and permafrost, are a significant carbon sink, sequestering carbon in the form of peat and soil organic matter. Peatlands, in particular, are the planet's largest terrestrial carbon sink, storing about 30% of the world's soil carbon (footnote 4).

[32] M. Ubierna, C. Diez Santos, and S. Mercier-Blais. 2022. Water Security and Climate Change: Hydropower Reservoir Greenhouse Gas Emissions. In A. K. Biswas and C. Tortajada, eds. *Water Security Under Climate Change*. Singapore: Springer.

[33] O. Edenhofer et al., eds. 2014. Climate Change 2014: *Mitigation of Climate Change—Working Group III Contribution to the Fifth Assessment Report of the Intergovernmental Panel on Climate Change*. Cambridge, United Kingdom and New York: Cambridge University Press. Sections 7.5.1 and 7.8.1.

[34] J. Poore and T. Nemecek. 2018. Reducing food's environmental impacts through producers and consumers. *Science*. 360 (6392). pp. 987–992.

[35] H. Ritchie and M. Roser. 2022. Emissions by sector. https://ourworldindata.org/emissions-by-sector (accessed 20 December 2022).

[36] A. Chalak et al. 2016. The global economic and regulatory determinants of household food waste generation: A cross-country analysis. *Waste Management*. 48. pp. 418–422.

[37] J. Gustavsson et al. 2011. *Global Food Losses and Food Waste: Extent, Causes and Prevention*. Rome: Food and Agriculture Organization of the United Nations.

[38] P. Smith et al. 2014. Agriculture, Forestry and Other Land Use. In O. Edenhofer et al., eds. *Climate Change 2014: Mitigation of Climate Change—Working Group III Contribution to the Fifth Assessment Report of the Intergovernmental Panel on Climate Change*. Cambridge, United Kingdom and New York: Cambridge University Press.

When wetlands are drained, disturbed, or converted to other land uses, much of the carbon stored is released into the atmosphere in the form of CO_2 and CH_4 and dissolved organic carbon, and nitrogen enter water courses, contributing to GHG emissions (footnote 4).

More than 85% of global wetlands that existed before the industrial revolution had been lost by 2000, the majority of which occurred in the 20th century. Wetlands continue to be lost at a rate three times faster than forest losses.[39] Wetland losses have been especially acute in Asia and the Pacific, accounting for an estimated 40% of global wetland loss.[40]

In Asia and the Pacific, land use conversion has been driven by various human activities, including agriculture, urbanization, and infrastructure development. The Mekong Delta in Viet Nam, for example, is among the world's most productive rice-growing regions, but at the cost of conversion of its wetlands, leading to the loss of much of its carbon storage capacity. The Sundarbans mangrove forest, which spans the Bay of Bengal coastlines of Bangladesh and East India, faces multiple threats, including conversion of mangroves to shrimp farms and other aquaculture industries, leading to a loss of carbon storage capacity. Similarly, wetlands and forests throughout Asia and the Pacific have been converted to agriculture, including animal husbandry. With rising income levels, demand for meat and fish products has steadily increased throughout Asia and the Pacific and are projected to continue to grow at a robust rate in the future.[41] Meat production has the highest GHG emissions of all foods, with beef emitting 100 kilograms of CO_2 equivalent per kilogram (beef herd)—up to 20 times that of rice per kilogram (footnote 34). Meat production is also remarkably water intensive. A kilogram of beef requires about 15,000 liters of water to produce.[42] Other meat products—such as pork and chicken—similarly produce higher GHG emissions than nonmeat agriculture products.

Land use change and forestry accounts for about 7% of global emissions, as shown in Figure 3.

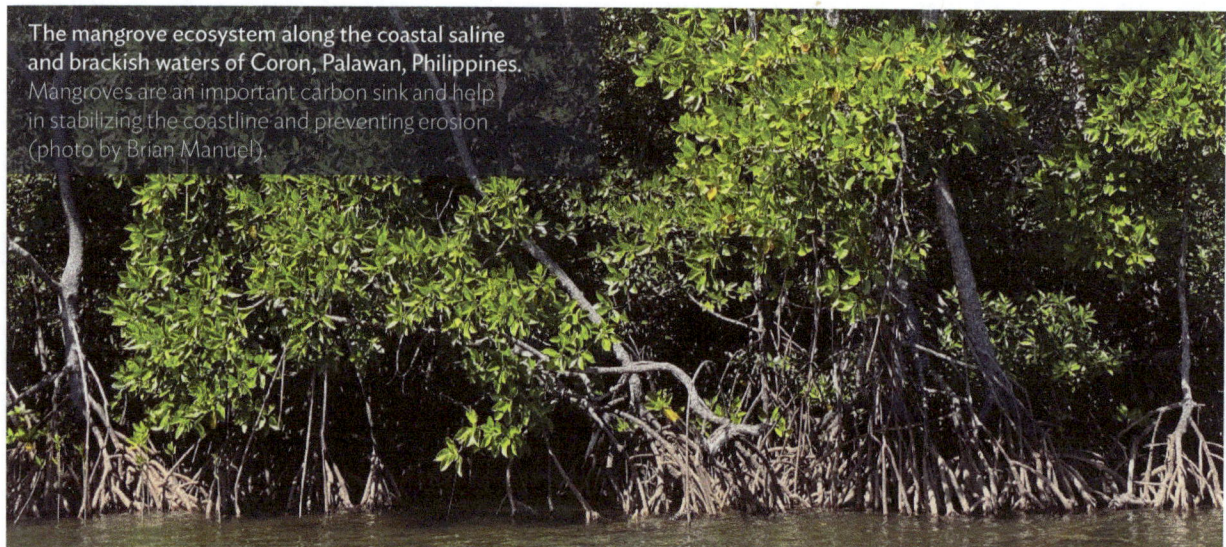

The mangrove ecosystem along the coastal saline and brackish waters of Coron, Palawan, Philippines. Mangroves are an important carbon sink and help in stabilizing the coastline and preventing erosion (photo by Brian Manuel).

39   Ramsar Convention on Wetlands. 2018. *Resolution XIII.13: Restoration of degraded peatlands to mitigate and adapt to climate change and enhance biodiversity and disaster risk reduction.* Enacted at the 13th Meeting of the Conference of the Contracting Parties to the Ramsar Convention on Wetlands: Wetlands for a Sustainable Urban Future. Dubai, United Arab Emirates. 21–29 October; and United Nations Educational, Scientific and Cultural Organization and UN-Water. 2020. *The United Nations World Water Development Report 2020:* Water and Climate Change. Paris.

40   N. Anisha et al. (footnote 4), citing S. Hu et al. 2017. Global wetlands: Potential distribution, wetland loss, and status. *Science of the Total Environment.* 586. pp. 319–327.

41   Statista. Meat—Asia: Revenue. (accessed 18 December 2022).

42   Calculations based on data in Water Footprint Calculator. Food's Big Water Footprint.

BOX 4

Water in the Development of the Asia and Pacific Region

Water is central to the Asia and Pacific region's economic development and sustainable future as water resources profoundly impact the region's land use, agriculture, energy, and other vital infrastructure. Climate change impacts causing droughts, floods, loss of glaciers and snow pack, salt water intrusion into freshwater aquifers, and other impacts threatening freshwater supply are already affecting Asia and the Pacific, and these impacts are expected to intensify under continued climate change. The Asian Water Development Outlook produced by the Asian Development Bank every 3–4 years assesses national water security across the Asia and the Pacific, with a focus on five key dimensions: rural, economic, urban, environmental, and water-related disaster.

Water scarcity. In Asia and the Pacific, water scarcity impacts agriculture, and will ultimately threaten industry and urban water resources, putting at risk decades of progress in economic development. Water shortage has been exacerbated by rapid urbanization. According to a global study of water risk conducted by the WWF, more than 31 cities in the People's Republic of China (PRC); 30 cities in India; 12 cities in Indonesia; 8 cities in Pakistan; and additional cities in Bangladesh, Kazakhstan, Malaysia, Mongolia, the Philippines, the Republic of Korea, Singapore, and Viet Nam will all face severe water scarcity by 2050.[a]

Water contamination. Fresh water in Asia and the Pacific, particularly in coastal areas and low-lying deltas, is vulnerable to seawater intrusion during extreme wave and storm surge events, and, in the longer run, by rising sea levels. Fresh water in rural areas is vulnerable to pollution from mismanagement of sanitation, agricultural runoff, and salination due to over-irrigation.

The Pacific islands, especially atoll islands, are perhaps the most at risk. Only 55% of Pacific island populations are able to access basic drinking water and only 30% have sanitation services—the lowest rate in the world.[b]

Precipitation events and flooding. Climate-related extreme precipitation events will overwhelm urban infrastructure, causing more frequent flooding and flood-related disasters, including disabling the power grid and overloading the sewerage network. Extreme precipitation events will impact rural areas, inundating agricultural land and damaging crops. Droughts will destroy crops and render land dry and unusable. These events can cause long-term contamination of freshwater sources and permanently change agricultural patterns. In areas where water resources are overstressed or mismanaged, these impacts will be magnified.

Decarbonizing infrastructure for water, water treatment, sanitation, hygiene, irrigation, and water resources management presents an opportunity to upgrade these systems for greater resilience to extreme weather events.

Forest and wetland vulnerability. Forests, specifically rainforests, are not only important carbon sinks that sequester carbon dioxide from the atmosphere, but they also protect and stabilize groundwater resources and form an important component of the local/global hydrological cycle. More than a dozen developing countries in Asia and the Pacific host significant rainforests, which are at risk because of development and logging interests, both legal and illegal. Losing these forests will result in increased greenhouse gas emissions and risks undermining water security.

continued on next page

Box 4   *continued*

Wetlands serve as filters for pollutants and sediment, and store carbon in their plant communities and soil.[c] Relative to forests, wetlands such as peatlands have higher carbon density and, in general, flooded inland marshes and coastal canopy wetlands are less susceptible to fire and other threats, although peatlands are at risk of combustion as water tables decline.[d] The world's wetlands contribute to storing a significant portion of organic carbon for both terrigenous landscapes and oceans.[e]

Mountain glacial systems. Increasing temperatures have accelerated the melting of glacial systems of the Hindu Kush Himalaya range and the Tian Shan range, which are home to some of the world's largest high-altitude glacial systems. Populations of the southern PRC, India, Central Asia, and Southeast Asia depend on glacial runoff for their water. The Hindu Kush Himalayan range alone feeds the Ganges-Brahmaputra, Indus, Mekong, Yangtze, and Yellow rivers, providing water for drinking, irrigation, and other uses for about 1.5 billion people in India, Pakistan, the PRC, and Southeast Asia.[f] The Tian Shan range supplies water to roughly another 100 million people throughout Central Asia and northwest PRC.

Beyond water for drinking and agricultural uses, Asia also depends on these glacial systems for hydropower electricity essential to powering its economic development.

Oceans and seas. Oceans play a crucial role in the global carbon cycle by absorbing and storing carbon dioxide from the atmosphere, acting as a natural carbon sink. However, pollution can negatively impact this important function in several ways. Pollution, including ocean acidification, nutrient pollution, plastic pollution, and oil spills, can disrupt marine ecosystems and hinder their capacity to absorb and store carbon dioxide. By reducing the oceans' role as a carbon sink, pollution exacerbates the challenges of climate change and highlights the importance of addressing these environmental issues to preserve the health of our oceans.

---

[a] Global Water Intelligence. 2022. *Mapping Water's Carbon Footprint: Our net zero future hinges on wastewater.* Oxford. p. 5.

[b] WWF Water Risk Filter. Lists of Cities at risk (Scenarios) 2020. https://wwfindia-my.sharepoint.com/:x:/g/personal/kchaudhary_wwfindia_net/EQJsIxVi76hHn0HX6Ns_onQBXKlGTVuscnFjmoF-MF1zsw?rtime=UEzENmvn2kg (accessed 26 December 2022).

[c] M. Wilson et al. 2022. *Political Economy of Water Management and Community Perceptions in the Pacific Island Countries.* San Francisco: The Asia Foundation.

[d] United States Environmental Protection Agency. 2022. *Why are Wetlands Important?* https://www.epa.gov/wetlands/why-are-wetlands-important.

[e] D. L. A. Gaveau et al. 2014. Major atmospheric emissions from peat fires in Southeast Asia during non-drought years: evidence from the 2013 Sumatran fires. *Scientific Reports.* 4.

[f] B. Kayranli et al. 2010. Carbon Storage and Fluxes within Freshwater Wetlands: a Critical Review. *Wetlands.* 30 (1). pp. 111–124; and C. M. Duarte, J. J. Middelburg, and N. Caraco. 2005. Major role of marine vegetation on the oceanic carbon cycle. *Biogeosciences.* 2 (1). pp. 1–8.

[g] National Research Council. 2012. *Himalayan Glaciers: Climate Change, Water Resources, and Water Security.* Washington, DC: The National Academies Press.

Source: Asian Development Bank.

Mangrove management training. The training was for members of San Miguel Unity for Progress in San Balangiga, Samar, Philippines the majority of whom are women (photo by Eric Sales).

3 Water Sector Decarbonization Measures

The water sector presents numerous opportunities for decarbonization. Because water is integrated in all aspects of life and economic development, these opportunities span a wide range of water infrastructure and water-use activities.

This section covers water decarbonization opportunities in the following areas:

(i)   Water supply
(ii) Sanitation, and hygiene
(iii) Water resources management: energy and water storage
(iv) Irrigated agriculture
(v)   Land use and forestry resource management

These opportunities interconnect with each other at the nexus of water use, conservation, and land preservation. For example, enhancing water efficiency and promoting water conservation by residential, business, and industrial users greatly reduces the volumes of water required to be processed by water supply and wastewater infrastructure, and thereby permanently reduces GHG emissions. Similarly, integrated planning of urban development and agricultural and other land uses with water conservation and decarbonization as twin goals can drive efforts to reduce land conversion of wetlands and other carbon-storing lands.

3.1 Water Supply

Water supply is energy intensive and, by extension, carbon intensive. Electricity accounts for virtually all water abstraction and distribution emissions, because of operation of pumps and processing equipment throughout the treatment process (footnote 21). Opportunities for decarbonizing the water supply subsector therefore requires primarily decarbonizing its supply of electricity and energy efficiency, such as high-efficiency pumps. Siting an elevated water treatment plant can reduce pumping and provide energy-saving benefits, but it requires careful planning, consideration of topography, and proper design of the distribution network to ensure gravity flow.

Seawater desalination facilities are as much as three times more carbon intensive than water reclamation and treatment facilities, most of which is due to of energy consumption, although chemicals used in the process also contribute to emissions.[43] For water reuse technologies, which impose higher standards for effluents, and thus require additional processing, energy consumption is the dominant driver of emissions, contributing about 68%–92% of a facility's carbon footprint (footnote 43).

Energy Source

For water processing facilities that can host their own power generation, solar and wind offer the lowest GHG emissions and have the least water impact (Box 2, footnote b). They are also resilient against water scarcity. Solar panels can even be constructed over water canals to take maximum advantage of available space, and to shelter them and reduce evaporation, preventing water loss.[44] Additionally, water can cool the panels making them more efficient.

For facilities that cannot generate their own electricity, reducing emissions of electricity generation will require sourcing non-fossil sources such as solar, wind, and mini-hydropower, thereby reducing their scope 2 emissions (Section 3.2). This can be achieved through reaching power purchase agreements with clean energy generators, or by acquiring renewable energy credits or carbon offsets for those emissions, provided such credits or offsets are available in the particular country.

Water Efficiency and Conservation

Enhancing water efficiency and conservation reduces the volume of water that must be produced or abstracted, and the volume of wastewater treated, thereby reducing energy consumed and by extension GHG emissions in both the water and energy sectors over the entire life cycle of a water asset. Water efficiency and conservation delivers multiple benefits, and because of its savings in both emissions and cost, should be among the first measures deployed. Importantly, these cost savings should be passed to consumers in a properly regulated market, an important consideration, especially in an economic development context.

Consumer-focused efficiency and conservation initiatives have long been the starting point in the energy sector for government. In the water sector, consumer-focused efficiency and conservation initiatives to reduce GHG emissions and reduce peak demand for electricity are alternatives to building more infrastructure, and can generate cost savings for consumers. Importantly, relative to energy sector conservation efforts, customer-focused water conservation programs have proven to be as or more cost-effective in reducing water use and GHG emissions.[45] As consumption expands, utilities source new water supply from increasingly distant and/or lower-quality sources, which require more energy and cost for pumping and treatment. Additional factors influence water system emissions intensity, including water system size, topography (including elevation for gravity transport), water source type, climate, tariff structure, and varying fuel mixes of power

---

43   P. K. Cornejo et al. 2014. Carbon footprint of water reuse and desalination: A review of greenhouse gas emissions and estimation tools. *Journal of Water Reuse and Desalination*. 4 (4). pp. 238–252; and X. Jia et al. 2019. Analyzing the Energy Consumption, GHG Emission, and Cost of Seawater Desalination in China. *Energies* 12 (3). p. 463.

44   B. McKuin et al. 2021. Energy and water co-benefits from covering canals with solar panels. *Nature Sustainability*. 4. pp. 609–617.

45   E. S. Spang, S. Manzor, and F. J. Loge. 2020. The cost-effectiveness of energy savings through water conservation: a utility-scale assessment. *Environmental Research Letters*. 15 (11).

generation reflected in the grid emissions factor.[46] As distance and/or treatment requirements increase, the embedded energy intensity, and therefore carbon intensity, of delivered water increases (footnote 45).

Water efficiency and conservation measures are an effective and low-cost way to reduce the carbon footprint and costs involved in energy usage. Examples of water efficiency and conservation measures include

- low-flow faucets;
- high-efficiency appliances such as dishwashers and washing machines;
- low water use and dry composting toilets and/or urinals;
- pressure control via regulators and variable speed drive pumps, saving water and energy;
- drip irrigation systems for agriculture;
- smart technologies for efficiency, including monitoring and sensors to detect water supply leaks;

- rate design to incentivize water-saving behavior; and
- storage and time-of-use pricing to shift consumption from peak to off-peak periods.

Reduction of the physical (real) losses of nonrevenue water will be an important strategy in water conservation efforts. Any water lost by a water utility in the process of abstracting, treating, and delivering water to the end consumer will translate into a financial loss for the utility, will generally incentivize excess usage, therefore contributing to GHG emissions without providing the intended benefit.

Water conservation programs can be bundled with electricity efficiency programs, such as increasing block tariff structures and time-of-use pricing incentivizing shifting consumption to off-peak periods, to enhance cost savings to consumers and reduce the cost of delivery of these programs.

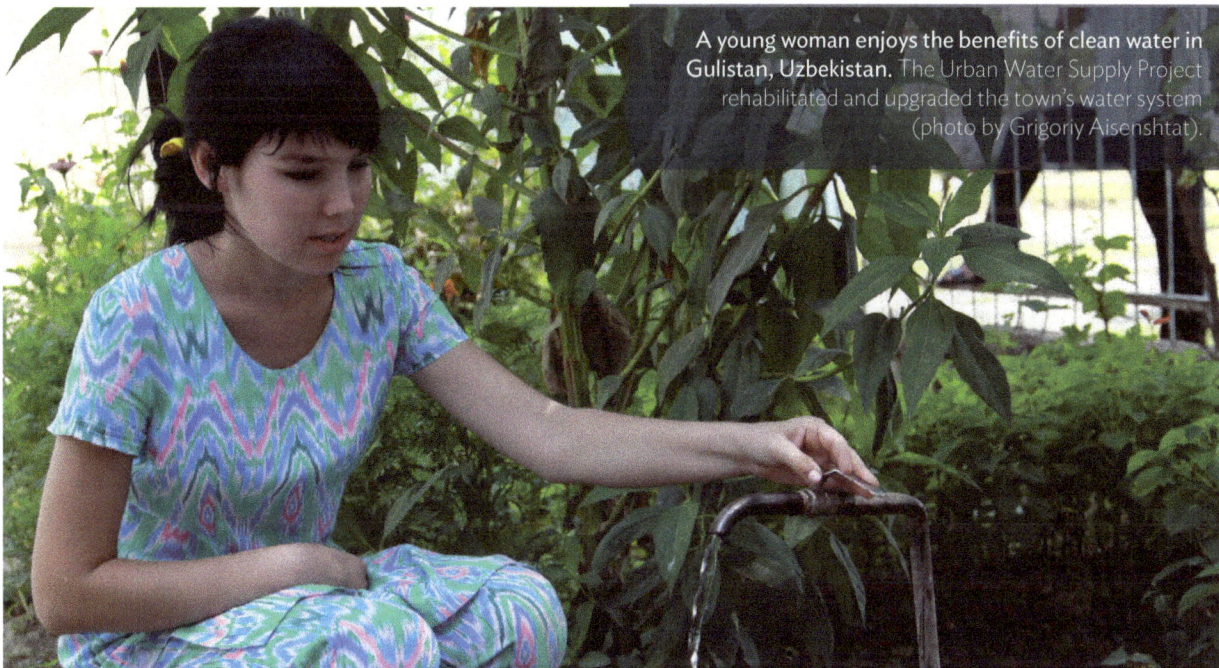

A young woman enjoys the benefits of clean water in Gulistan, Uzbekistan. The Urban Water Supply Project rehabilitated and upgraded the town's water system (photo by Grigoriy Aisenshtat).

---

[46] R. B. Sowby and S. J. Burian. 2018. Statistical model and benchmarking procedure for energy use by U.S. public water systems. *Journal of Sustainable Water in the Built Environment.* 4 (4); and R. B. Sowby and A. Capener. 2022. Reducing carbon emissions through water conservation: An analysis of 10 major U.S. cities. *Energy Nexus.* 7.

Standards for water and energy efficiency programs can also be deployed that will ensure only water and energy products are available in the country. Water and electricity pricing that promotes conservation should be a policy priority.

Countries in Asia and the Pacific have already adopted some water efficiency and conservation measures that serve as a starting point for further action. The PRC introduced an urban water pricing reform in the early 2000s that adopted increasing block rate pricing, which has been shown to reduce residential water demand modestly.[47] India promotes rainwater harvesting, recharging of underground aquifers, construction of check dams, and other water conservation measures.[48] The Energy Efficiency and Conservation Sub-sector Network and its Plan of Action for Energy Cooperation, 2016–2025 of the Association of Southeast Asian Nations (ASEAN) incorporates water conservation as part of its programs of promoting policy priorities and shared standards among its member countries.[49] ADB has long promoted water utility efficiency through regulation, tariff reform, and infrastructure improvements for the entire Asia and the Pacific region.[50]

3.2 Wastewater, Sanitation, and Hygiene

The wastewater, sanitation, and hygiene subsector presents perhaps the greatest opportunities for GHG emissions reduction in the water sector in Asia and the Pacific.

This section focuses on modern wastewater treatment facilities, as these systems are increasingly common in Asia and the Pacific.

This is especially true for rapidly urbanizing regions, where water and wastewater infrastructure has expanded significantly and opportunities for carbon reduction in these facilities are increasing along with the volume of wastewater processed and sludge produced. To illustrate, in the PRC, the wastewater infrastructure has expanded along with the population living in urban areas and the provincial per capita gross domestic product. During the PRC's past 2 decades of rapid growth, the PRC expanded wastewater treatment capacity, and the production of wastewater sludge grew at an average annual rate of 44.9% from 2005 to 2010, and 7.3% from 2010 to 2019.[51] In addition, section 3.2.7 also covers decentralized dry sanitation systems.

Reducing emissions during the wastewater treatment phase should begin with water efficiency and conservation measures, described in section 3.1. Reducing the volume of wastewater that must be treated reduces the energy consumed and the resulting GHG emissions in both the water and energy sectors associated with the treatment process.

This section presents opportunities to decarbonize WWTP operations, organized as follows:

(i)   Optimizing WWTP operations
(ii)   Sludge disposal
(iii) Sludge to energy
(iv)   Sludge to minerals recovery
(v)   Carbon capture, use, and storage
(vi)   Thermal recovery
(vii) Decentralized sanitation systems: dry sanitation systems

[47]   B. Zhang, K. H. Fang, and K. A. Baerenklau. 2017. Have Chinese water pricing reforms reduced urban residential water demand? *Water Resources Research*. 53 (6). pp. 5057–5069.

[48]   V. K. Swamy. 2022. 10 NGOs working for water conservation in India. *Give Blog*. 13 May.

[49]   ASEAN Centre for Energy. 2017. *Best Practices in Energy Efficiency and Conservation: ASEAN Energy Awards*. Jakarta.

[50]   ADB. 2020. *Asian Water Development Outlook 2020: Advancing Water Security across Asia and the Pacific*. Manila.

[51]   L. Wei et al. 2020. Development, current state and future trends of sludge management in China: Based on exploratory data and CO_2-equivalent emissions analysis. *Environment International*. 144.

In analyzing emissions reduction opportunities, this section adopts the Greenhouse Gas Protocol's scopes 1, 2, and 3 framework[52] in which the WWTP is the scope 1 facility. Scope 1 emissions are those occurring within the fence of the WWTP, such as CO_2, CH_4, and N_2O emissions from aerobic and anaerobic treatment units, and CO_2 emissions from burning of non-biogenic CH_4 to produce electricity. Scope 2 emissions are those related to electricity purchased by these facilities from third-party generators for their operations. Electricity carbon emissions are generally dependent on the grid emissions factor where the WWTP is located. Because of reliance on coal-fired power generation, emissions factors are generally higher in Asia and the Pacific than in other regions. Scope 3 emissions are all other emissions of non-electricity suppliers of goods and services and consumers, such as off-site sludge disposal or incineration performed by third parties; emissions embedded in buildings and agriculture; and, more broadly, emissions driven by land use changes such as conversion of wetlands associated with the WWTP.

Fundamentally, new approaches will require rethinking wastewater infrastructure and developing new technologies and processes to be incorporated into future designs. The analysis in sections 3.2.1–3.2.6 evaluates decarbonization opportunities that can be performed on existing WWTPs, making changes to existing operations. For additional description of WWTP emissions, see the technical appendix to this guidance note.

Optimizing Wastewater Treatment Plant Operations

Wastewater treatment emits GHGs throughout the process:

- During the aeration process, scope 1 CO_2 is emitted. Also, aeration accounts for significant scope 2 emissions, typically demanding as much as 50%–90% of electricity consumption in wastewater treatment operations.[53]
- The denitrification stage release of N_2O is of particular concern because it is a potent GHG with a global warming potential of 273 times that of CO_2 over both 20-year and 100-year periods.[54]
- At present, separation of N_2O presents technical and economic challenges, making it a priority for decarbonization research efforts.
- During anaerobic treatment, CH_4 is produced. If vented, fossil-derived CH_4 has a global warming potential of 29.8 times that of CO_2 over a 100-year period and 82.5 times over a 20-year period (footnote 54). CO_2 can also be used to generate power for WWTP operations.

Operations can be varied to reduce GHG emissions, as well as to achieve various goals, such as optimizing for effluent volume or purity, economics, and energy consumption. Depending on the goals, variations can include (i) controlling inflow quantity, composition, and mixing; (ii) adopting advanced treatment such as secondary clarifying treatment; (iii) ensuring the absence of secondary treatment; (iv) optimizing aeration levels and duration; (v) providing chemical treatment of sludge; and (vi) using inorganic precipitates.

[52] Greenhouse Gas Protocol.
[53] J. Drewnowski et al. 2019. Aeration Process in Bioreactors as the Main Energy Consumer in a Wastewater Treatment Plant. Review of Solutions and Methods of Process Optimization. *Processes*. 7 (5); W. Tomczak and M. Gryta. 2022. Energy-Efficient AnMBRs Technology for Treatment of Wastewaters: A Review. *Energies*. 15 (14); and Z. Bao, S. Sun, and D. Sun. 2016. Assessment of greenhouse gas emission from A/O and SBR wastewater treatment plants in Beijing, China. *International Biodeterioration and Biodegradation*. 108. pp. 108–114.
[54] P. Forster et al. 2021. The Earth's Energy Budget, Climate Feedbacks and Climate Sensitivity. In V. Masson-Delmotte et al., eds. Climate Change 2021: *The Physical Science Basis—Working Group I Contribution to the Sixth Assessment Report of the Intergovernmental Panel on Climate Change*. Cambridge, United Kingdom and New York: Cambridge University Press. pp. 923–1054, Table 7.15.

Depending on the goals sought to be achieved, process changes may be synergistic with certain other goals and may compete with others. For example, process variations that reduce GHG emission may involve tradeoffs in effluent quality or energy consumption, or potentially partially offsetting emissions.

Depending on the adopted goals of the WWTP operator, process changes or improvements can be made that reduce GHG emissions. Examples of efficiency measures and other upgrades that can reduce GHG emissions include the following:

- **Reducing and/or capturing methane emissions from the anaerobic digester.** Scope 1 CH_4 emissions may include venting of CH_4 and fugitive emissions (leakage) from the anaerobic digester and pipes, and incomplete combustion of biogas in flaring equipment. CH4 emissions are determined by the quantity of biogas collected at the digester outlet, the percentage of biogas that is vented or leaked from the digester and pipes,[55] and flare efficiency. Eliminating venting through process changes (either installing flares or utilizing biogas for productive purposes) and plant improvements that ensure leaks are eliminated will reduce scope 1 emissions.
- **Reducing energy consumption emissions.** Scope 2 indirect emissions are emissions from the generation of electricity consumed in the wastewater and sludge treatment process. Efficiency upgrades to plant equipment that reduce energy consumption, or changes in the supply of electricity such as installation of combined heat and power (CHP) units or renewables will reduce emissions.
- **Reducing chemical consumption.** Scope 3 indirect emissions include, for example, emissions embedded in the chemicals used to operate the WWTP. Reduced reliance on chemicals or replacement with less carbon-intensive chemicals could reduce wastewater scope 3 emissions.

More ambitious GHG reductions efforts will focus on more fundamental changes to plant operations. Possible approaches being explored include changes to the secondary treatment process, focusing on aeration and nitrification-denitrification:

- As aeration can account for as much as 50%–90% of electricity consumption in wastewater treatment operations (footnote 53), reducing, replacing, or eliminating aeration in the secondary treatment process with anaerobic or other processes can significantly reduce overall GHG emissions.
- Developing alternatives to conventional nitrification-denitrification processes, which require electricity, require chemical additives, and produce potent N_2O GHG emissions.

In addition to aeration, reducing sludge production is another priority area involving fundamental changes to plant operations that could reduce GHG emissions as well as achieve more environmentally sustainable and economic outcomes. These efforts may focus on the following:

- Processes that reduce sludge production by replacing traditional sludge treatments such as dewatering, drying, stabilization, and anaerobic digestion with new processes.

Sludge Disposal

Sewage sludge describes the "solid, semi-solid, or liquid residue generated during the treatment of domestic sewage in a treatment works."[56] "Biosolids" is an industry term that describes the material produced after sludge has been treated to meet standards for landfill disposal, incineration, or beneficial use, such as in land applications.[57]

---

[55] G. Zhao et al. 2021. Greenhouse Gas Emission Mitigation of Large-Scale Wastewater Treatment Plants: Optimization of Sludge Treatment and Disposal. *Polish Journal of Environmental Studies*. 30 (1). p. 955–964.

[56] United States Environmental Protection Agency. 1999. *Standards for the Use or Disposal of Sewage Sludge*. 40 CFR 1.503.9. Washington, DC.

[57] United States Environmental Protection Agency. Basic Information about Biosolids.

Wastewater treatment plant in Apia, Samoa. The Sanitation and Drainage Project aims to improve public health, environmental quality, and surface and groundwater quality, and to reduce frequency of flooding in specific low-lying areas (photo by Luis Enrique Ascui).

Sewage sludge commonly contains significant amounts of pollutants (including heavy metals, viruses, bacteria, and other pathogens), requiring treatment before it is safely released into the environment without risk to human health and the environment.[58]

Reducing the mass of sludge or biosolids for disposal would reduce WWTP emissions and lead to more environmentally and financially sustainable outcomes. GHG emissions associated with sludge and biosolids treatment and disposal, mostly from energy consumption, can account for 40%–75% of total wastewater treatment emissions.[59] Sludge and biosolids disposal are increasingly regulated because of health and environmental considerations. Reducing the mass of solids produced can significantly reduce operations costs because their treatment and disposal can account for as much as 50% of

WWTP operating costs.[60] Reductions in sludge production will also alleviate demand for landfills near populated centers, which has become a severe constraint to urban development and sanitation in developing countries in Asia and the Pacific, and can contribute to reductions in air and soil pollution caused by landfill mismanagement (footnote 51).

Reducing sludge production could be achieved by acting on the variables that affect the quantity of sludge produced in the treatment process. These include the composition and volume of influent wastewater, the specific treatment processes employed, and applications of increasing levels of treatment.

The quantity of influent wastewater generally increases with population, regulation, wastewater infrastructure coverage, and treatment

---

58 K. Fijalkowski et al. 2017. The presence of contaminations in sewage sludge: The current situation. *Journal of Environmental Management.* 203 (3). pp. 1126–1136.

59 S. Brown, N. Beecher, and A. Carpenter. 2010. Calculator tool for determining greenhouse gas emissions for biosolids processing and end use. *International Journal of Environmental Science and Technology.* 44. pp. 9509–9515.

60 City of Detroit Water and Sewerage Department. 2014. *Biosolids Dryer Facility Project Plan May 2014.* Detroit; and G. Zhao et al. 2021. Greenhouse Gas Emission Mitigation of Large-Scale Wastewater Treatment Plants: Optimization of Sludge Treatment and Disposal. *Polish Journal of Environmental Studies.* 30 (1). pp. 955–964.

effectiveness.[61] Conservation and efficiency measures for water, as well as chemicals used in agricultural, industrial, and residential applications, can help reduce the amount of sludge production by reducing wastewater and water pollution loads.

Reducing sludge disposal mass could also be achieved through conversion of wet sludge to biofuels and incinerating dry sludge in waste-to-energy applications, both reducing scopes 1 and 2 facility emissions by avoiding additional sludge processing emissions and costs, and avoiding scope 3 emissions resulting from disposed sludge and biosolids.[62]

The specific emissions pathways associated with sludge and biosolids disposal are outlined in the following list. Beyond reducing the mass of sludge and biosolids produced, additional measures for reducing emissions specific to each pathway are identified as follows:

- **Landfill — methane emissions.** Sludge disposed in landfills emits CH_4 in volumes determined by the mass of dry sludge and the CH_4 conversion factor based on the degradable organic content of the sludge, the type of disposal site (footnote 55), and environmental factors such as temperature.
- **Biosolids land fertilization — nitrous oxide emissions.** N_2O emission volumes are determined by the mass of dried sludge, the proportion of N_2O and other nitrogen compounds in the sludge, and environmental conditions. N_2O emissions spike following land application and after wet–dry cycles and freeze–thaw cycles. Emissions can be reduced by timing application to early spring to mitigate summertime peaks or by using a nitrification inhibitor.[63]

- **Incineration — nitrous oxide emissions.** N_2O emission volumes are determined by the mass of dried sludge, the proportion of N_2O and other nitrogen compounds in the sludge (footnote 55), and the operating conditions and efficiency of the incineration facility.
- **Transport — vehicle carbon dioxide emissions.** Transportation of sludge typically via diesel trucks creates CO_2 emissions. Conversion to electric trucks (when available) could reduce emissions depending on grid emission factors and the source of electricity.

Sludge can be used for fertilization. In addition to pollutants, wastewater influent also typically contains valuable nutrients, particularly nitrogen, phosphorus, and organics, that can be recycled as sludge for fertilizer after undergoing the appropriate treatments. Cost-effective recovery of these nutrients creates the possibility of displacing fossil-fuel fertilizer production and application.[64]

Sludge to Energy

Sludge Incineration for Energy

Some sludge can be incinerated to recover energy and generate electricity. Incineration may require auxiliary fuel for sustained combustion at desired temperatures. The auxiliary fuel consumption should be insignificant and no greater in emissions intensity to the grid emissions factor.

Biogas Recovery for Combined Heat and Power

WWTPs can generate power or CHP for use within the plant, or for sale to third parties. Avoided emissions are determined based on the relative emissions of the plant's generation

[61]   D.-J. Lee et al. 2014. *Sludge Management, Handbook of Environment and Waste Management, Land and Groundwater Pollution Control.* Singapore: World Scientific. pp. 149–176.

[62]   T. E. Seiple, A. M. Coleman, and R. L. Skaggs. 2017. Municipal wastewater sludge as a sustainable bioresource in the United States. *Journal of Environmental Management.* 197. pp. 673–680.

[63]   D. Hunt et al. 2019. Year-Round N₂O Emissions from Long-Term Applications of Whole and Separated Liquid Dairy Slurry on a Perennial Grass Sward and Strategies for Mitigation. *Frontiers in Sustainable Food Systems.* 3.

[64]   D. Puyol et al. 2017. Resource Recovery from Wastewater by Biological Technologies: Opportunities, Challenges, and Prospects. *Frontiers in Microbiology.* 7; and Y. V. Nancharaiah, S. V. Mohan, and P. N. L. Lens. 2016. Recent advances in nutrient removal and recovery in biological and bioelectrochemical systems. *Bioresource Technology.* 215. pp. 173–185.

unit compared to the electricity grid. Captured biogas from the WWTP's anaerobic digester can be used in CHP units, producing higher thermal efficiencies compared to single power or heat processes. CHP systems can achieve system efficiencies of more than 80% in WWTPs.[65]

The electricity generated can significantly reduce emissions and plant operating costs. Every 1 million gallons (i.e., 3.8 million liters) of wastewater inflow per day can produce enough biogas in an anaerobic digester to produce 2.4 million British thermal units per day of thermal energy, which in a CHP system can produce 26 kilowatts of electric capacity or 624 kilowatt-hours of electricity (footnote 65). The electricity generated could be used in plants and other operations, displacing the use of grid electricity partly derived from fossil fuels.

Sludge to Liquid Biofuels

Wastewater and sludge contain carbon resources that can be utilized as feedstock to produce biofuels. Biofuels sourced from organic wastes can reduce GHG emissions on a life-cycle basis by displacing traditional fossil fuels, thereby enhancing their value over traditional fuels. Economies of scale can be achieved by sourcing sludge from a WWTP, animal husbandry operations, and food waste. Biofuels produced through reuse of organic wastes present opportunities to produce higher-value products than biogas from wastewater.[66]

The conversion efficiency, GHG reductions, and economics of sludge to biofuels depend on the design of the specific facility as well as the composition and consistency of influent wastewater.

Primary and secondary treated wastewater and sludge are both candidates as feedstocks for biofuels. Following treatment, the sludge undergoes a mechanical thickening dewatering stage that increases the solids concentration of the sludge to 15%–40%.[67]

Conversion efficiency greatly depends on the composition of the feedstock sludge. Generally, conversion efficiency of biochemical components adheres to the following hierarchy: lipids > protein > carbohydrates. Mixtures of proteins and carbohydrates also have higher yields than both components separately, thus the following hierarchy applies: mixture > proteins > carbohydrates (footnote 67).

Importantly, primary wastewater sludge has been observed to produce a biocrude yield of 37.3% with a corresponding high heating value of 37.8 megajoules per kilogram, while secondary wastewater sludge produces a biocrude yield of 24.8% and has a high heating value of 34.8 megajoules per kilogram (footnote 67).

The closeness in heating values between primary and secondary treated sludge suggests the opportunity to avoid the secondary treatment step, saving the cost of processing, and avoiding CO_2 emissions because of electricity consumption from aeration and N_2O emissions from the denitrification process that occurs during this step.

---

[65]   United States Environmental Protection Agency. 2011. *Opportunities for CHP at Wastewater Treatment Facilities: Market Analysis and Lessons from the Field.* Washington, DC.

[66]   S. Y. Li et al. 2021. Techno-economic uncertainty analysis of wet waste-to-biocrude via hydrothermal liquefaction. *Applied Energy.* 283.

[67]   R. L. Skaggs et al. 2018. Waste-to-Energy biofuel production potential for selected feedstocks in the conterminous United States. *Renewable and Sustainable Energy Reviews.* 82. pp. 2640–2651.

Sludge to Minerals Recovery

Sludge may also be recycled as a feedstock for building materials, eliminating disposal costs and decreasing demand on landfill sites. Recycling sludge into building materials can be combined with sludge incineration for electricity generation, using sludge ash and sintered residue, further enhancing revenues from monetizing sludge.

Sewage sludge contains components usable as raw material for construction aggregates, ceramic materials, cement, and brick. Usable minerals include silicon dioxide (10%–25% concentration), aluminum oxide (5%–10%), and calcium oxide (10%–30%); silicon dioxide increases in sludge ash after incineration to 25%–50% concentration, aluminum oxide to 10%–20%, and calcium oxide to 15%–30%. Sludge can substitute up to 15% of traditional feedstocks in cements and 20% in the fabrication of bricks, ceramic materials, and lightweight aggregates, while maintaining the quality of the final products.[68]

Wastewater treatment in the pulp and paper industry presents unique opportunities for sludge recovery. India, Indonesia, Japan, and the PRC are among the top 10 paper-producing countries.[69] Lime sludge generation volumes in pulp and paper manufacturing place it among the most polluting industries. Chemical recovery systems, which are standard in the industry, dissolve chemicals in water in their molten stage to form a liquor that must be processed to dispose of the water and sludge, which is generally landfilled.[70] Recovered paper sludge ash and alkali recovery paper sludge are being developed as an additive for higher-value calcium silicate board materials, used as construction and insulation panels. Concentrations of up to 30% paper sludge ash or 20% alkali recovery paper sludge can improve the board's flexural strength and reduce its thermal conductivity, thereby enhancing its effectiveness as an insulator, as well as improving its heat stability.[71]

Sewage sludge may even present economically viable opportunities for recovery of rare earth lanthanides, critical minerals (including scandium and yttrium), and other precious metals.[72]

Carbon Capture, Use, and Storage

WWTPs produce CO_2 emissions as a result of oxidation of fossil residues in sludge, as well as additional CO_2 emissions from incineration. WWTPs could employ carbon capture equipment to separate the carbon and resell it for reuse. Biogas production may require separation of CO_2 to produce gas adhering to applicable standards if sold for third-party use. Research and development of electrochemical separation or conversion of CO_2 to other separable and recoverable compounds such as CH_4 are being advanced, which may be more economical for WWTPs compared to traditional carbon capture approaches.[73]

Possible reuse options for captured CO_2 include concrete production, food and beverage industry uses, and as an injectant for oil production. Reused carbon embedded in products like cement are removed from the atmosphere and stored securely for the life of that product. Alternatively, instead

---

68   Z. Chang et al. 2020. *Valorization of sewage sludge in the fabrication of construction and building materials:* A review. Resources, Conservation and Recycling. 154 (1). pp. 130–134.

69   Statista. *Leading pulp for paper producing countries worldwide in 2021* (accessed 25 December 2022).

70   M. Chen et al. 2019. Recycling of paper sludge powder for achieving sustainable and energy-saving building materials. *Construction and Building Materials*. 229.

71   P. Vashistha et al. 2019. Valorization of paper mill lime sludge via application in building construction materials: A review. *Construction and Building Materials*. 211. pp. 371–382.

72   M. B. Folgueras et al. 2018. An Overview of Rare Earth Elements in Sewage Sludges and Their Ashes. *Proceedings*. 2; R. Kaegi et al. 2021. Quantification of Individual Rare Earth Elements from Industrial Sources in Sewage Sludge. *Water Research X*. 11; and F. Suanon et al. 2017. Assessment of the occurrence, spatiotemporal variations and geoaccumulation of fifty-two inorganic elements in sewage sludge: A sludge management revisit. *Scientific Reports*. 7.

73   J. M. Sonawane, Z. J. Ren, and D. Pant. 2022. Carbon valorization using the microbial electrochemical technology platform. In Z. J. Ren and K. Pagilla, eds. *Pathways to Water Sector Decarbonization, Carbon Capture and Utilization*. London: IWA Publishing.

of reuse, carbon can also be stored by injection underground in geologic formations capable of permanent containment.[74]

Depending on whether the organic material producing CO_2 in wastewater and sludge is deemed anthropogenic or biogenic for the purposes of carbon accounting or government regulation, carbon removal by carbon capture, use, and storage may be deemed to achieve net negative removals from the atmosphere. Net negative removals can produce higher-revenue value depending on incentives for biogenic GHG removals (footnote 74).

Thermal Recovery

Energy recovery from wastewater also presents opportunities to reduce net carbon emissions. Wastewater in sewers, treatment plants, and post-treatment reservoirs are potential thermal sources for water source heat pumps. Thermal energy is usable for heating and cooling without undergoing energy conversion and associated conversion losses. The potential thermal energy of wastewater is significantly greater than the chemical energy extracted through biogas recovery by an estimated 6–8 times.[75]

Thermal recovery systems have been deployed by WWTPs and sewage infrastructure in Australia, Japan, and the PRC as well as in Europe and North America (footnote 75). Thermal recovery through district geothermal presents the possibility of new business models for water and wastewater treatment utilities to provide heating, cooling, and decarbonization services.

Decentralized Sanitation Systems: Dry Sanitation Systems

For rural and peri-urban areas, decentralized sanitation systems are essential to delivering sanitation services on a cost-effective basis. Open defecation, wet and dry pit latrines,

and septic tanks are a potential source of soil contamination, disease, and GHG emissions.

Alternatives to centralized sanitation systems and septic tanks include dry sanitation systems that rely on natural processes such as composting and dehydration to convert human waste into a safe and usable form, reducing GHG emissions.

Dry sanitation systems offer several advantages relative to centralized sanitation systems, wet sanitation systems and septic tanks, specifically for rural and peri-urban applications.

Dry sanitation systems can significantly reduce water use relative to centralized sanitation systems. Modern flush toilets typically use from 1 to 2 gallons of water, compared to older models that used as much as 6 gallons of water each cycle, resulting in significant household water consumption for flushing toilets alone. In contrast, dry sanitation systems do not use water to transport human waste, eliminating the need for water in the sanitation process. This can result in significant water savings, particularly in areas where water is scarce or expensive.

Dry sanitation systems can also reduce GHG emissions. Flush sanitation systems rely on energy-intensive water abstraction, transport, and, after use, wastewater collection and treatment, which consume large amounts of electricity, typically generated by fossil fuels, contributing to GHG emissions. In contrast, dry sanitation systems do not require WWTPs, eliminating the need for energy-intensive processes. The natural processes used in dry sanitation systems, such as composting and dehydration, are powered by the sun and wind, reducing the carbon footprint of the sanitation process.

Beyond energy savings and avoided water use, dry sanitation systems can produce nutrient-rich compost that can replace chemical fertilizers.

---

[74]   A. C. Jones and A. J. Lawson. 2022. Carbon Capture and Sequestration (CCS) in the United States. *Congressional Research Service Report*. No. R44902. Washington, DC: Congressional Research Service.

[75]   X. Hao et al. 2019. Energy recovery from wastewater: Heat over organics. *Water Research*. 161. pp. 74–77.

Synthetic chemical fertilizers are energy intensive to produce and generate significant GHG emissions.[76] By replacing chemical fertilizers with compost, dry sanitation systems further reduce GHG emissions.

3.3 Water Resources Management: Energy and Water Storage

Water resources management, especially for hydroelectric power generation and reservoir applications, presents significant opportunities for decarbonizing the water sector.

Several methods are available to reduce GHG emissions from reservoirs and mitigate their impact on the environment.

Siting and land use practices concerning reservoirs are an essential method to reduce reservoir GHG emissions. Siting reservoirs should minimize impacts relative to pre-project conditions, such as not constructing reservoirs on land that requires

clearing of forests or destruction of indigenous aquatic habitats through reservoir inundation.[77] Siting should also facilitate land use practices that aim to prevent or reduce the amount of fertilizer and other organic matter that can enter reservoirs. These methods include locating siting reservoirs in upland areas with lower temperature microclimates that reduce aquatic plant growth;[78] and promoting agriculture practices near reservoirs such as conservation tillage, crop rotation, and cover cropping that can reduce soil erosion and avoid fertilizer runoff into reservoirs, as well as sequester carbon in soils (footnote 8). Depriving the reservoir of organic nutrients and depriving these nutrients of an optimal growing environment reduces the amount of algae and other plant matter that grows in the reservoirs, in turn reducing the extent of organic matter decay under anaerobic conditions and thereby the amount of CH_4 released.

Another method is to aerate and de-stratify (mix) reservoir water, which helps increase oxygen levels and reduce thermal stratification in the reservoir, thereby scaling down the anaerobic decomposition process, and reducing the amount

Water resources management. The Gumain Dam in Bataan, Philippines (photo by Doods Tanabe).

---

[76]   S. Menegat, A. Ledo, and R. Tirado. 2022. Greenhouse gas emissions from global production and use of nitrogen synthetic fertilisers in agriculture. *Scientific Reports*. 12.

[77]   ADB. 2023. Guidance Note on Large Hydropower Plants. Unpublished.

[78]   R. M. Almeida et al. 2019. Reducing greenhouse gas emissions of Amazon hydropower with strategic dam planning. *Nature Communications*. 10.

of CH_4 produced.[79] These methods require electricity, and energy usage should be considered in the overall emissions calculation. To avoid energy-related emissions, the electricity required to operate aeration, mixing, and pump equipment can be powered by renewable energy sources, like solar and wind, located on top of or near the reservoirs.

Another method is to construct hydropower facilities to avoid or reduce the reservoir area. Alternatively, by changing the reservoir operating levels, it is possible to reduce the size of the shallow littoral area hence reducing the GHG emissions. Run-of-river hydropower does not generate significant GHG emissions since the impounded water volumes are usually very small.

When designing and constructing dam facilities (including the dam body, spillway, and any power generation substations), reducing the emissions during construction and those embedded in construction materials, and maintaining the infrastructure can further reduce GHG emissions on a life cycle basis. In particular, the steel and concrete used in constructing dams embed significant GHG emissions. Methods to reduce these emissions include using concrete that stores carbon, designing infrastructure to reduce the amount of concrete required, and incorporating alternative materials into dam construction (as these materials are developed) and green cement.[80] Another opportunity is by designing the intake above the thermocline, which separates the bottom layer from the top layer, where more mixing occurs. By doing this, the amount of CH_4 emissions that would be released through degassing can be reduced.[81]

Technological innovation will advance other methods to decarbonize reservoirs. One such method involves installing reservoir CH_4 capture mechanisms that capture and convert the CH_4 bubbling from reservoirs into electricity, a technology that is being demonstrated in Brazil.[82]

3.4 Irrigated Agriculture

Water efficiency and conservation efforts must be joined by all sectors of the economy, including households, industry, and agriculture, to significantly reduce water sector energy consumption and GHG emissions.

Agriculture, especially, must be the focus of water conservation programs.[83] Agriculture accounts for an estimated 70% of global water consumption.[84] As the dominant consumer of water, the volume of water used by agriculture enables highly competitive energy savings in upstream operations of abstraction, conveyance, treatment, and distribution, even though agricultural water conservation may not contribute to energy savings from avoided wastewater collection and treatment, depending upon the particular system (footnote 45).

Changing farming techniques to reduce water consumption and emissions is critical, especially for rice production. For Asia, rice farming followed by soy farming are the among the most GHG intensive because of CO_2, CH_4, and N_2O emissions. Possible solutions include substituting these crops with high-value crops that require less water and/or are less GHG intensive. Also, where possible, drip irrigation can be employed as a conservation measure. Section 3.1 describes general conservation approaches that can also be applied to agriculture.

79   F. Guérin et al. 2015. Effect of sporadic destratification, seasonal overturn and artificial mixing on CH4 emissions at the surface of a subtropical hydroelectric reservoir. *Biogeosciences*. 13 (12). pp. 3647–3663.

80   J. Jia et al. 2016. The Cemented Material Dam: A New, Environmentally Friendly Type of Dam. *Engineering*. 2 (4). pp. 490–497.

81   Y. T. Prairie et al. 2017. *The GHG Reservoir (G-Res) Tool: Technical documentation. Version 3.2 (19 December 2022)*. United Nations Educational, Scientific and Cultural Organization Chair in Global Environmental Change and International Hydropower Association.

82   Global Innovation Lab for Climate Finance. Reservoir Methane Capture Mechanism.

83   D. M. Ruiz et al. 2020. Turning off the tap: Common domestic water conservation actions insufficient to alleviate drought in the United States of America. *PLoS ONE* 15 (3).

84   OECD. Water and agriculture.

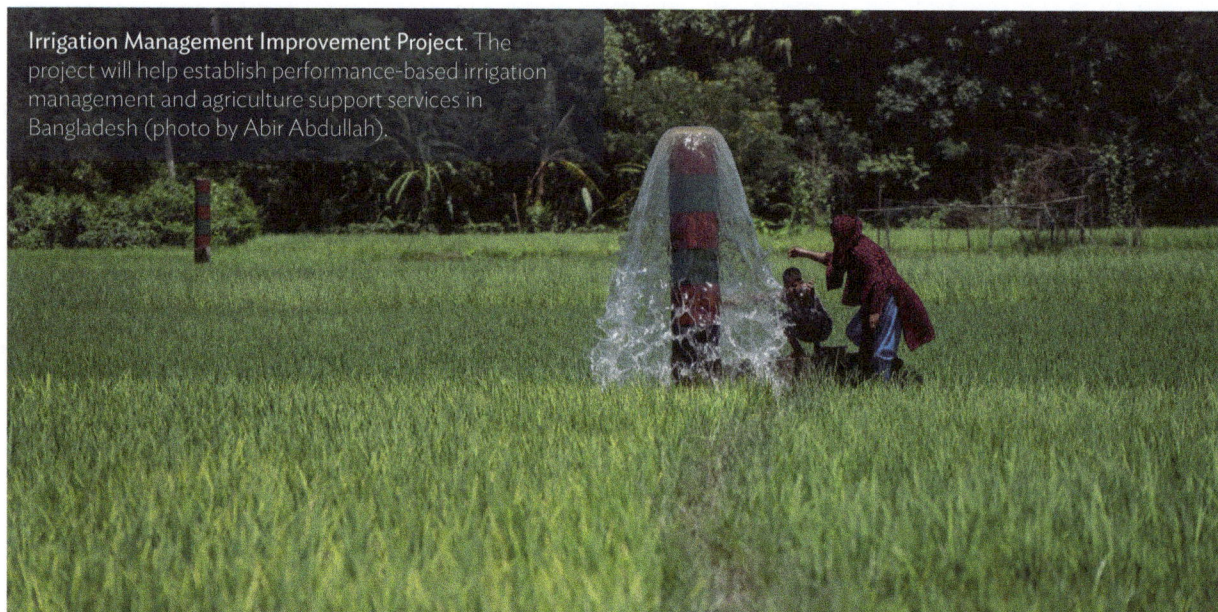

Irrigation Management Improvement Project. The project will help establish performance-based irrigation management and agriculture support services in Bangladesh (photo by Abir Abdullah).

For rice, an approach known as the "system of rice intensification" (SRI) uses far less water than traditional rice paddy farming. Instead of flooding crops, the SRI keeps soil moist through intermittent flooding and drying, and transplants young seedlings with greater spacing, enabling yields to be maintained or even improved.[85] The SRI's practice of intermittent flooding significantly reduces GHG emissions by limiting the anaerobic conditions that produce CH_4 emissions and reducing groundwater demand by 1.5–2.5 times per hectare, thereby reducing fossil energy used for pumping. Reductions in synthetic fertilizers and higher yields relative to conventional rice paddy farming further contribute to the SRI reducing GHG emissions by as much as 64% per kilogram of rice produced.[86]

The SRI has been adopted in more than 20 countries throughout Asia and the Pacific, and more worldwide.[87] The SRI and other techniques to enhance agriculture and irrigation efficiency, as well as improvements to transport and storage infrastructure, will help prevent food loss, which also contributes to greater water consumption and GHG emissions (footnote 37).

These water-saving measures can be combined with other GHG-saving measures like low- or no-till farming. Substituting organic fertilizers, such as those derived from organic wastes from WWTPs described in section 3.2.2, can further reduce the carbon intensity of farming by reducing chemical inputs. Reducing the chemicals used in farming, in turn, reduces chemical runoff into water sources which causes eutrophication and reduces the processing requirements for water treatment facilities. Reduced water processing reduces water sector GHG emissions.

An increasingly significant driver of food-related water consumption and emissions is changing patterns of food waste at the consumption stage, another source of water loss. Food waste at the consumption stage is increasingly occurring in industrializing Asia to levels approaching those in developed countries, driven by rising household income and urbanization (footnote 36).

---

85   Strategies for survival. 2017. *Nature Plants*. 3 (907).
86   A. Gathorne-Hardy et al. 2016. System of Rice Intensification provides environmental and economic gains but at the expense of social sustainability — A multidisciplinary analysis in India. *Agricultural Systems*. 143. pp. 159–168.
87   SRI International Network and Resources Center. SRI Information by Country.

3.5 Land Use and Forestry Resources Management

Urbanization and development have come at a heavy cost to our natural environment. Many of the services originally provided by our natural ecosystem are being replaced with human-made infrastructure when development expands. When natural systems are replaced with human-made infrastructure, GHG emissions increase—this is especially true for water infrastructure.

Development patterns matter greatly for water sector GHG emissions. Forests, especially rainforests, are not only important carbon sinks that remove CO_2 from the atmosphere, but they also protect and stabilize groundwater

resources, and they help generate hydrological cycles. Wetlands serve as filters for pollutants and sediment, and they store carbon in their plant communities and soil (Box 3, footnote c) unities and soil. Forests and wetland geographies (such as peatlands, mangroves, and eelgrass) connect the ecologies of agriculture, watersheds, and river habitats. These are the earth's natural sanitation and wastewater systems that help ensure clean and safe water, and protect ecosystems and water cycles.

A holistic approach to decarbonizing the water sector in Asia and the Pacific requires protecting forests and wetlands. Nature-based solutions can be superior in performance and cost to human-made infrastructure. To the extent that their carbon sink capacity can be preserved and possibly enhanced through integrated

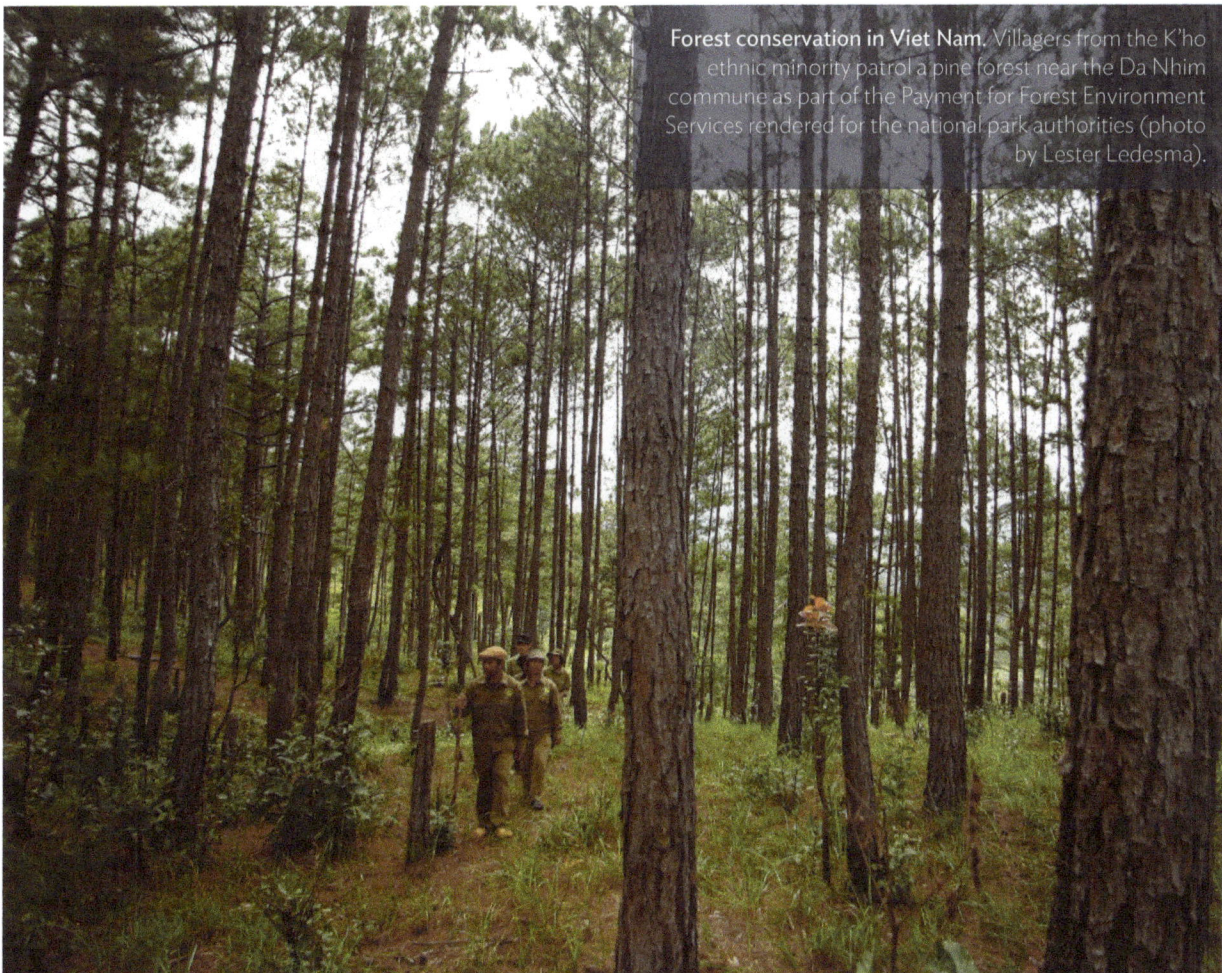

Forest conservation in Viet Nam. Villagers from the K'ho ethnic minority patrol a pine forest near the Da Nhim commune as part of the Payment for Forest Environment Services rendered for the national park authorities (photo by Lester Ledesma).

planning, decarbonization can be scaled through preventative actions, which are more effective than restorative measures.[88] A preventative approach will help protect these habitats for both their environmental and economic development services.

Measures to reduce water sector emissions should include protecting the forests and wetlands that provide water and other environmental services. Land use and development planning should integrate water resources in decision-making and government approval and permitting. Adopting integrated water resources management with climate resilience and GHG emissions reduction as objectives helps balance development objectives, with the imperative of protecting water resources that provide a service in support of communities. Integrated water planning can also enhance resilience. Cities located close to water bodies could either be at heightened risk because of climate events or, if properly planned and developed, could develop safer and more water-secure futures.[89]

In general, these criteria imply a more comprehensive and rigorous approach to development planning that must consider the entire water cycle, including natural resources, chemicals and energy inputs, and resulting wastewater pollution.[90] Figure 4 illustrates the components of the urban water cycle, suggestive of the potential synergies that can be achieved through integrated planning.

Within the framework of integrated water resources management, land planning and development approval and permitting could include water-related measures, specifically

- maintaining and regenerating forest, peatlands, mangroves, eelgrass, and other vegetation;
- incorporating urban design elements that minimize life-cycle water and carbon impacts;
- promoting wastewater treatment in areas without treatment, and developing infrastructure using best available technologies for meeting environmental objectives, including minimizing GHG emissions;
- designing water systems to maximize opportunities to integrate mini- and micro-hydro and pumped storage and solar (traditional or floating);
- enabling dual water systems for potable and non-potable water that can exploit rainwater harvesting and use partially treated wastewater where appropriate;
- operating water systems to optimize efficient use of water resources and infrastructure together with energy infrastructure through information technology approaches and real-time coordination of processes;
- ensuring affordable, sustainable, and diverse sources of water supply;
- ensuring compliance with GHG emissions mandates; and
- requiring water projections and cooperation on an urban and water basin basis.

---

[88] J. B. Gallagher, K. Zhang, and C. H. Chuan. 2022. A Re-evaluation of Wetland Carbon Sink Mitigation Concepts and Measurements: A Diagenetic Solution. *Wetlands.* 42.

[89] A. Bahri. 2012. Integrated Urban Water Management. *Global Water Partnership Technical Committee Background Papers.* No. 16. Stockholm: Global Water Partnership.

[90] C. A. Peña-Guzmán et al. 2017. Urban Water Cycle Simulation/Management Models: A Review. *Water.* 9 (4).

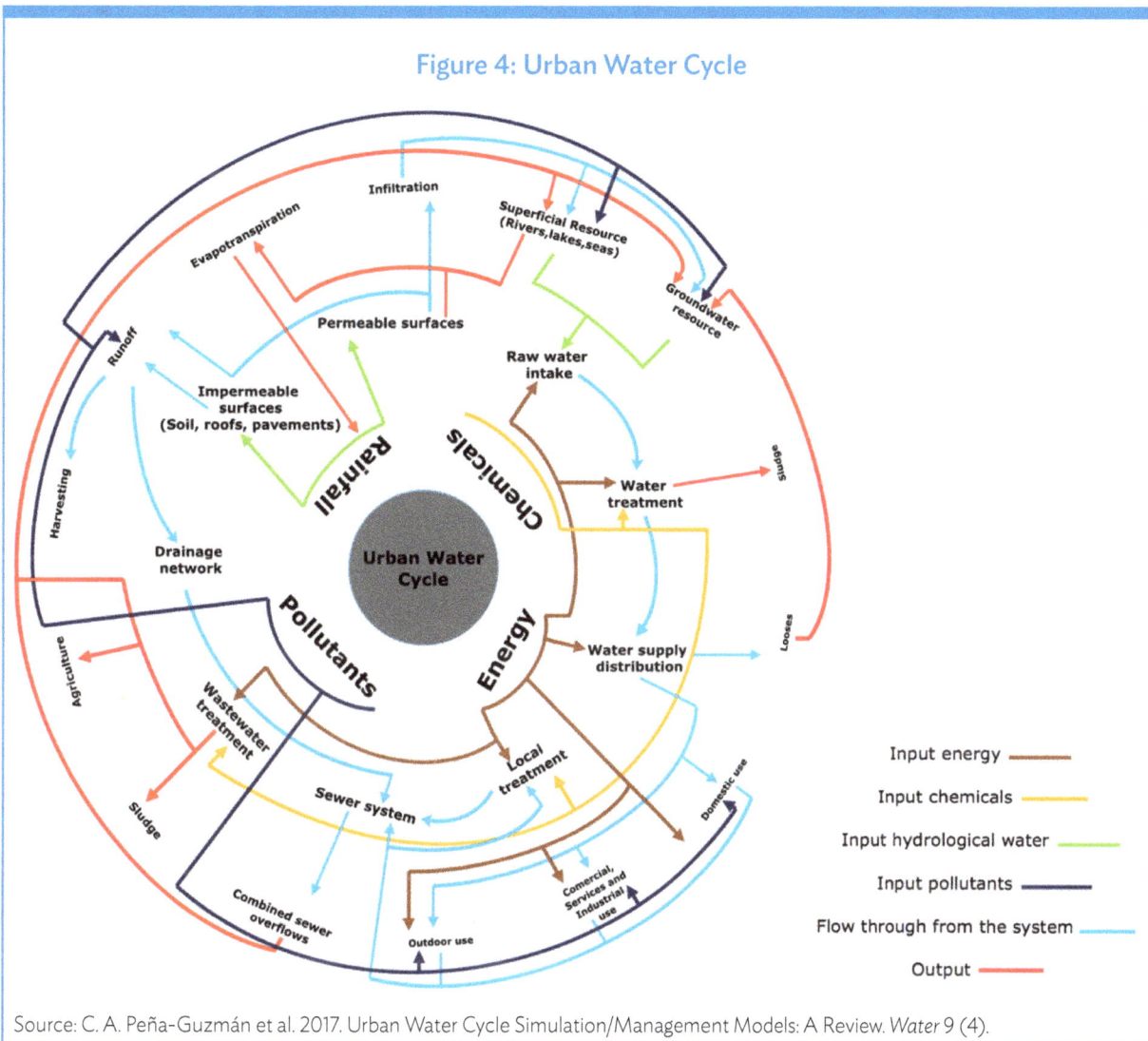

Figure 4: Urban Water Cycle

Source: C. A. Peña-Guzmán et al. 2017. Urban Water Cycle Simulation/Management Models: A Review. *Water* 9 (4).

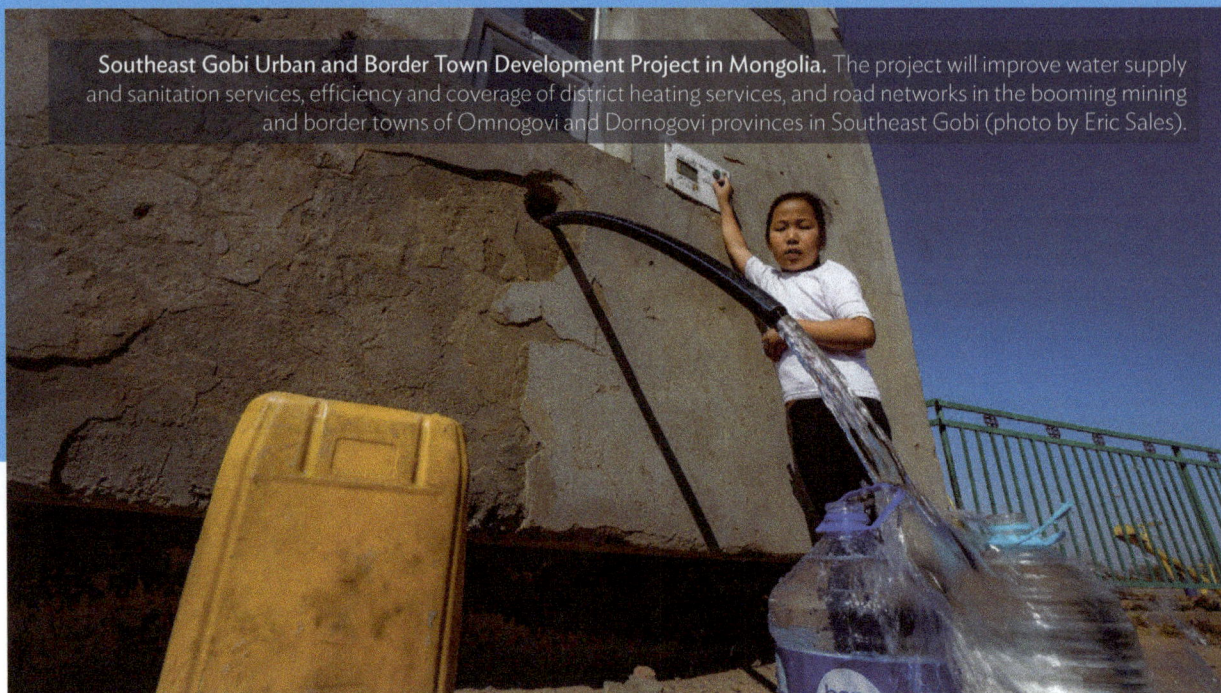

Southeast Gobi Urban and Border Town Development Project in Mongolia. The project will improve water supply and sanitation services, efficiency and coverage of district heating services, and road networks in the booming mining and border towns of Omnogovi and Dornogovi provinces in Southeast Gobi (photo by Eric Sales).

4 Policies and Tools to Support Water Sector Decarbonization

Because the water sector is integral and connected to all aspects of the natural environment and economic development, this section on policies to support water decarbonization covers policies of general application and subsector-specific policies that promote decarbonization as well as water resilience for the following subsectors: water supply; wastewater, sanitation and hygiene; water resources management; irrigated agriculture; and land use and forestry. This section also examines how international institutions and regional approaches can promote and even be essential to achieving water decarbonization.

4.1 General Water Policies and Tools

Carbon Price for Water

To decarbonize the water sector in the most economically efficient way possible, establishing a price for carbon embedded in the provision of water would help internalize the costs of GHG emissions.

A price for carbon could be system-wide, reflecting an average or policy estimate of water emissions, or could be differentiated by function, technology, source, and other characteristics.

Clean water extraction and distribution, desalination, and wastewater treatment all have unique carbon emissions profiles, and within each a wide range of variation because of multiple factors, including regulatory standards, distance to source, water system size, topography, water quality, climate, and grid emissions factors (footnote 46).

Whether a generalized system-wide price or differentiated price of carbon approach is adopted, the pricing of carbon emissions is a powerful lever to incentivize utilities and consumers to incorporate GHG emission considerations into infrastructure investment and consumption decisions.

Carbon Footprinting

Managing GHG emissions in the water sector requires measuring these emissions.

Water sector GHG emissions have not been well monitored and reported on a sector-wide basis. Rather, water sector and water-relevant emissions are often embedded within GHG emissions from the energy sector and land use, land use change, and forestry subsectors. The lack of specific tracking for the water sector has produced gaps in data because water has been treated as embedded in other (sub)sectors, and thus neglected as its own sector in carbon inventories, estimates, and solutions.

The water sector requires greater data visibility to facilitate decarbonization and the adoption of carbon credits, to improve the certainty of emissions estimates, and to identify synergies and tradeoffs among decarbonization and improvements in delivering water, sanitation, and hygiene (WASH) services to underserved populations, and in adapting water infrastructure for greater resilience.

Accordingly, a system of water emissions footprinting incorporating ongoing monitoring, reporting, and (eventually) verification should be adopted by countries in Asia and the Pacific. Water carbon footprinting standards that are appropriate to the conditions prevailing in the region and the capabilities of the governments and their utilities should be developed and adopted, gradually increasing in rigor with experience and technical capacity.

Measurement of GHG emissions should be performed on a full life-cycle assessment basis for scopes 1, 2, and 3, as this information will inform government policy and approvals, and public and private investment decisions.

Tools for Estimating Water Sector Greenhouse Gas Emissions

Several tools are available to measure GHG emissions in the water sector, including

- the Energy Performance and Carbon Emissions Assessment and Monitoring Tool (ECAM);
- the GHG Reservoir (G-Res) Tool;[91]
- the Screening Tool for Energy Evaluation of Projects (STEEP); and
- other methodologies, primarily those developed under the Clean Development Mechanism (CDM).

ECAM[92] is a software tool designed to measure GHG emissions and energy consumption on a system-wide level, including for water supply (water abstraction, treatment, and distribution) and sanitation (wastewater collection, treatment, and on-site sanitation). It uses a combination of process modeling, emission factor databases, and activity data to estimate emissions of CH_4, N_2O, and CO_2. ECAM provides detailed information on the sources of GHG emissions and identifies opportunities for reducing emissions. It is a useful tool for water utilities to identify sources of emissions and assess the effectiveness of mitigation strategies.

---

[91]   G-res Tool.
[92]   ECAM.

The G-Res Tool is a software tool designed to assess the net GHG footprint from freshwater reservoirs. The G-Res Tool's operating principles include

- the GHG footprint of the landscape prior to impoundment;
- the particular environmental setting of each reservoir (climatic, geographic, edaphic, and hydrologic);
- the temporal evolution of the GHG emissions over the lifetime of the reservoir;
- displaced GHG emissions, i.e., emissions that would have occurred somewhere else in the aquatic network regardless of the presence of a reservoir;
- emissions increasing the net GHG emission impact of the reservoir, but that are the result of release of nutrients and organic matter by human activity occurring upstream of or within the reservoir.
- emissions from construction.

STEEP[93] is a is a free Microsoft Excel-based tool and reference guide that can be used to make system assessments. STEEP can also be used to identify potential areas for energy use savings in existing or planned water supply and wastewater facility projects.

The Clean Development Mechanism and the Paris Agreement

The CDM is a market-based mechanism under the Kyoto Protocol of the United Nations Framework Convention on Climate Change (UNFCCC) that enables industrialized countries to meet their emissions reduction targets by investing in emission reduction projects in developing countries. The CDM developed several methodologies for measuring GHG emissions in the water sector through a rigorous process of scientific and technical review. These methodologies are informed by IPCC scientific findings, including the 2006 IPCC Guidelines for National Greenhouse Gas Inventories (as updated and supplemented), and have been approved by the UNFCCC. The methodologies cover a range of water-related activities, such as wastewater treatment, water supply, and irrigation. These methodologies provide a standardized approach to measuring GHG emissions, which ensures consistency and transparency in the measurement process.

These methodologies were adopted and further evolved in voluntary carbon markets globally. The Paris Agreement Article 6 cooperative mechanisms, using CDM methodologies as the basis, provide opportunities for scaling up carbon emissions trading in the form of internationally transferred mitigation outcomes. Kyoto Protocol flexible mechanisms, voluntary markets, and the evolving Article 6 trading markets offer opportunities to translate GHG emission reductions into a revenue stream that can help developing countries realize mitigation as well as adaptation outcomes. Carbon crediting mechanisms can support decarbonization of the water sector by contributing to the investment costs. Opportunities for carbon credit development in water sector entities include the following, which are covered in this guidance note:

- water efficiency and conservation in the household, industry, and agriculture sectors;
- CH_4 avoidance and CH_4 capture and utilization;
- energy efficiency and electrification, in particular eliminating diesel-driven pumps and replacing less-efficient pump equipment with high-efficiency electrical pumps;
- renewable energy for power supply of water infrastructure;
- thermal energy recovery; and
- additional measures such as improved water resource and land use management, including preservation of forests, wetlands, peatlands, mangroves, and eelgrass, as well as restoration of these lands.

---

93   ADB. 2021. *Screening Tool for Energy Evaluation of Projects: A Reference Guide for Assessing Water Supply and Wastewater Treatment Systems.* Manila.

4.1.5 Water and Food Conservation Education

Water conservation and efficiency should be promoted through education. Water conservation and efficiency in all its forms should further include avoiding food waste.

For water conservation and efficiency, water tariff design that incentivizes conservation and efficiency is a foundational step. Tariff reform should be coupled with education to ensure the incentives are understood and effective.

Avoiding food waste is more difficult to incentivize because food consumption cannot be easily monitored and subjected to economic penalties for overconsumption.

A key element in both water and food conservation efforts is public education, preferably starting at a young age, socialized through the educational system and broadcast to the general public through the media and public-serving utilities.

Eliminating Subsidies to Fossil Energy, Agriculture, and Forestry

Because GHG emissions in the water supply and wastewater treatment subsectors in particular are determined significantly by energy source, agricultural practices, and land use decisions, subsidies to these non-water sectors contribute to water sector GHG emissions. Eliminating these subsidies is therefore critical to putting society on a more environmentally sustainable path, as it would reduce emissions from these sectors as well as from the water sector. Reducing these subsidies would also reduce government spending, making government operations more financially sustainable, and enabling governments to invest in other priorities.

Subsidies for the fossil energy subsector, which includes oil, gas, and coal, artificially lower the cost of fossil fuels, encouraging their consumption and hindering the growth of renewable energy sources. Subsidies to fossil fuel energy generation are significant. The International Monetary Fund estimates that global fossil fuel subsidies amounted to $5.9 trillion in 2020, or 6.5% of global gross domestic product.[94] By eliminating these subsidies, the true cost of fossil fuels will be reflected in their market price, making renewable energy sources more competitive. This shift will reduce the demand for fossil fuels and, subsequently, reduce GHG emissions.

Subsidies for the agriculture sector often include subsidized water use, enabling inefficiency and encouraging the use of synthetic fertilizers, which are themselves heavily subsidized, contributing to their overuse and GHG emissions.[95] Overfertilization, in which soil and water receive greater amounts of nitrogen and phosphates, produces nitrite emissions, a potent GHG. Furthermore, subsidies for large-scale monoculture farming practices encourage deforestation and the use of pesticides, both of which contribute to GHG emissions (footnote 95). The Organisation for Economic Co-operation and Development (OECD) estimates that distortive subsidies to the agriculture sector totaled $777 billion in 2021, more than one-third of which was provided by the PRC.[96] Eliminating these subsidies would discourage overfertilization and could encourage sustainable farming practices such as agroforestry, which can sequester carbon in the soil and reduce GHG emissions.

Subsidies for the forestry sector incentivize deforestation and unsustainable land practices.[97] Deforestation releases carbon stored in trees and soil, contributing to global GHG emissions.

---

[94] International Monetary Fund. 2021. Climate Change: Fossil Fuel Subsidies.
[95] Food and Agriculture Organization of the United Nations. 2020. *The State of Food and Agriculture 2020*. Rome.
[96] ADB. 2023. *Asia in the Global Transition to Net Zero: Asian Development Outlook 2023 Thematic Report*. Manila. p. 75.
[97] G. Kissinger. 2015. UN-REDD Programme Policy Brief—*Fiscal incentives for agricultural commodity production: Options to forge compatibility with REDD+*. New York: United Nations.

Monoculture forestry practices reduce biodiversity and have negative impacts on the environment, further contributing to GHG emissions. Monoculture farming of products like soy, beef, palm oil, and fast-growing timber for pulp and paper have driven destruction of critical tropical forests in Asia and elsewhere.[98] Eliminating subsidies for the forestry sector would incentivize more sustainable forestry practices, such as conservation and restoration of forests.

Eliminating subsidies for the fossil energy, agriculture, and forestry sectors would discourage unsustainable practices that contribute to GHG emissions and hinder the growth of renewable energy sources. By eliminating these subsidies, the true cost of these practices would be reflected in their market price, encouraging sustainable practices and reducing GHG emissions.

Private Sector Opportunities

The role of the private sector is essential in providing the required resources, financing, and know-how to foster the decarbonization of the water sector—with governments providing leadership, removing barriers, and supporting industry efforts through policies that mobilize markets to achieve environmental objectives.[99] Driven by demand from investors, customers, and employees, most of the world's largest listed companies have committed to bold climate actions. Pledges to achieve net-zero emissions have been issued by a number of high-profile companies.[100] The finance sector has equally stepped up its net-zero ambitions: more than 129 banks managing $74 trillion in assets (41% of total banking assets) have joined the Net-Zero Banking Alliance, which was convened by the United Nations in April 2021.[101] Specific opportunities where the private sector could play an active role in the water sector may include the following, among others:

(i)   Implementation of innovative technologies for carbon and CH_4 capture, energy efficiency improvements, efficient use and reuse of water resources, nonrevenue water reduction (especially physical losses), desalination, and climate-smart agriculture.

(ii)   Water treatment, distribution, sewerage, and sanitation services provided under concession agreements for water supply, sewerage, and sanitation services in a certain area and for a specified number of years, including activities such as:

- CH_4 capture and energy generation from WWTPs;
- nonrevenue water reduction;
- renewable energy generation and storage, e.g., solar and wind power at the utility's site and micro-hydropower and pumped storage, where appropriate;
- waste-to-energy and marketable products; and
- waste management and recycling.

(iii) WASH-related industries:

- energy efficient equipment production,
- energy service contracting,
- septic tank construction and emptying services, and
- water kiosks.

(iv)   Hydropower companies:

- optimization of water use and energy production,
- pumped storage schemes,
- floating solar and batteries, and
- CH_4 capturing from reservoir bodies to generate energy.

98   F. Seymour and J. Busch. 2016. *Why Forests? Why Now?: The Science, Economics, and Politics of Tropical Forests and Climate Change.* Washington, DC: Center for Global Development.
99   C. Hart. 2013. *The Private Sector and Climate Change: Scaling-up Private Sector Response to Climate Change.* London: Routledge.
100   South Pole, 2023. How should the private sector step up climate action. Position paper prepared by South Pole.
101   United Nations Environment Programme, 2023. Net-Zero Banking Alliance. Accessed on 9 May 2023.

(v) Farming and agriculture:

- sustainable agriculture and eco-friendly farming practices,
- water- and energy-efficient irrigation systems,
- carbon-neutral crop production practices, and
- CO_2 capturing from farmland by encouraging sustainable soil practices.

(vi) Private businesses, especially big water consumers such as beverage companies, mining companies, data centers, and textile industry businesses concerned about water supply, costs, and water quality, and their acceptance by local communities and the market, including instruments and activities such as the following:

- environmental, social, and governance ratings and energy efficiency ratings under which private companies make their own commitments to decrease their carbon footprints by adopting renewable energy, engaging in sustainable resource management, and setting examples for others in their industry;
- setting of science-based targets to reduce emissions within their operations and value chains, and compensating for their residual emissions through buying carbon credits;
- implementation of energy efficiency audits and disclosures;
- creation of long-term road maps to achieve net-zero emissions;
- a switch to renewable energy sources;
- development of low-carbon products and services;
- offsetting of emissions through carbon capture and storage or other mitigation measures;
- sourcing from local suppliers;
- waste conservation, sustainable packaging, and logistics;
- recycling of materials;
- environmental sustainable practices;

- investment in renewable energy (e.g., hydro, wind, solar, geothermal, ocean energy, and bioenergy);
- use of low-flow devices, reuse of gray water and effluent irrigation, and use of water recapture systems; and
- investment in clean energy and carbon capture and storage projects and technologies.

(vii) Carbon credits.

(viii) Green bonds as instruments to raise the needed financial capital.

Mechanisms such as the International Financial Reporting Standards S2 Climate-related Disclosures, which have been built on the recommendations of bodies such as the Financial Stability Board's Task Force on Climate-Related Financial Disclosures, could provide incentives to address concerns around insufficient disclosure of climate-related risks and opportunities.

4.2 Water Supply Policies

Tariff Design

Promoting water decarbonization must incorporate water supply tariff restructuring. Incentivizing water consumers to avoid water wastage, to encourage more efficient use of water, and to prioritize water use consistent with national economic priorities is an important first step to enlist household, industry, and agricultural consumers in decarbonizing the water sector.

For domestic water supply, best practice generally promotes tariff structures that promote water efficiency and conservation. These may include the adoption of inclining block rate pricing, which increases the volumetric charge to consumers for water as their consumption increases while providing lower rates for low-consumption households to meet essential needs. The PRC, for example, has adopted inclining block rate pricing for water supply, which has been shown to reduce residential water demand modestly (footnote 47). Other approaches include introducing seasonal and off-peak night rates that aim to reduce peak demand, and educating consumers about the rate structure and the potential savings in conserving water.

Water pricing should reflect the actual cost of water, including the cost of carbon emissions. Countries can calculate the GHG intensity of specific water resources as well as system-wide carbon intensity. Coupled with a cost of carbon, either based on a market-based carbon pricing mechanism such as carbon markets or a policy price of carbon (social cost of carbon), water rate design can internalize the price of carbon, which will incentivize conservation based in part on the carbon intensity of the water resource.

ADB has long promoted water utility efficiency through water tariff reform, regulatory reform, water infrastructure investment, and utility transformation for the entire Asia and the Pacific region (footnote 50).

Nonrevenue Water Loss

Nonrevenue water loss reduction (especially physical losses) should be an important element of water utility conservation efforts. Given the emission intensity of water abstraction, desalination, pumped distribution, and WWTP operations, any water lost by a water utility in the process of treating and delivering water to consumers is a loss financially to the utility and a loss to the environment in the form of GHG emissions over the entire water life cycle.

Efficiency Standards for Water Technologies and Products

A powerful tool to facilitate water supply subsector decarbonization is through the adoption of standards for water sector technologies, appliances, and household products. Standards encourage or mandate manufacturers to develop products with superior water-saving and emissions performance. These standards ensure that the options available in the marketplace

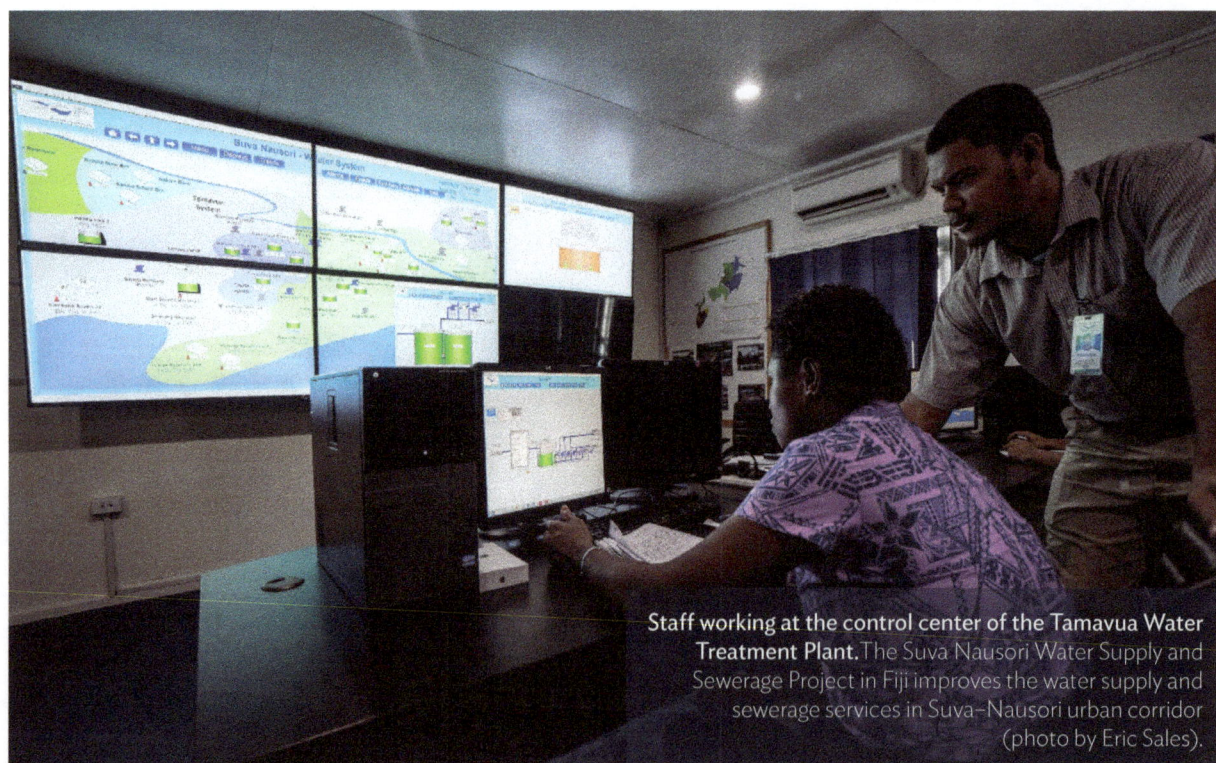

Staff working at the control center of the Tamavua Water Treatment Plant. The Suva Nausori Water Supply and Sewerage Project in Fiji improves the water supply and sewerage services in Suva-Nausori urban corridor (photo by Eric Sales).

comply with government mandates. They help educate the WASH subsector, the construction trades in the building sector, and consumers in making sustainable purchasing decisions.

Japan's Top Runner program, for example, has adopted standards for energy efficiency for dozens of product categories, requiring manufacturers to develop products to a higher performance standard, often resulting in their products being the leaders in their segment in energy and carbon efficiency. In doing so, Top Runner has not only produced results for the environment, but it has enabled leading companies to be more competitive in the marketplace (footnote 99). The PRC has adopted its own standards program modeled on Top Runner.

Harmonization of standards among countries generally is advantageous to driving down the cost of technology and encouraging trade. ASEAN, through its Energy Efficiency and Conservation Sub-Sector Network and its Plan of Action for Energy Cooperation, 2016–2025, incorporates water conservation as part of its programs of promoting policy priorities and shared standards for energy conservation among its member countries (footnote 49). These efforts, which already incorporate water, can be further expanded both in terms of water technologies and reach beyond ASEAN countries.

4.3 Sanitation and Hygiene Subsector Policies

The sanitation and hygiene subsector is the primary subsector responsible for ensuring that United Nations Sustainable Development Goal 6, aiming to ensure safe drinking water and provide safe sanitation for all, is achieved. Achieving Sustainable Development Goal 6's objectives involve providing underserved populations with

on-site sanitation or centralized wastewater infrastructure, which will increase the GHG emissions unless the WASH subsector is put on a path toward decarbonization. Further complicating this challenge, as populations urbanize and income levels increase, water consumption increases.[102]

Greater focus on the WASH subsector to ensure it decarbonizes in the face of these challenges must be prioritized. In addition to the general policies described in section 4.1, policies specific to the WASH subsector described in this section can help meet this challenge.

Financial Incentives

Financial incentives are essential for the wastewater subsector to invest in advanced infrastructure that not only safely manages wastewater, but also eliminates the carbon emitted in providing this essential service.

Decarbonization competes with a wide variety of regulatory and market forces directing the future of the wastewater treatment industry. Immediate priorities such as hygiene standards, expansion of infrastructure to meet demand and population growth, and water scarcity may take precedence over longer-term decarbonization objectives. Additionally, improving existing systems to eliminate per- and polyfluoroalkyl substances is rapidly emerging as yet another immediate priority for wastewater treatment facilities.[103] At the same time, rising energy costs and price volatility are significant drivers to decarbonize. Yet, the water sector in general remains under-resourced as competition for capital and sensitivity to consumer rates have led to underinvestment.

Technology incentives should be adopted in the context of national water planning, informed by integrated water resource planning that incorporates decarbonization and resilience as dual goals.

---

[102]   X. Ma et al. 2022. Environmental Science and Pollution Research. 29. pp. 4654–4667.
[103]   United States Environmental Protection Agency. PFAS Explained. https://www.epa.gov/pfas/pfas-explained.

A carbon price for water does not eliminate the need for financial incentives. Incentives such as new technology demonstration grants, concessionary interest rates (especially in a high-interest-rate environment), and tariff reform should all be considered, among other incentive options.

New technologies and approaches should be eligible for financial support, meaning that financial support should be performance driven. Performance-based incentives enable consideration of new approaches, such as considering decentralized water infrastructure, as well as better utilization of alternative sources of water, such as using gray water for toilets and other appropriate applications.[104]

Reducing Energy-Related Emissions in the Water Sector

Water infrastructure is carbon intensive because it is energy intensive. Electricity consumption accounts for virtually 100% of the emissions of water abstraction, treatment and network operations, and desalination processes. Electricity accounts for about 43% of the emissions of WWTPs (footnote 21). For water reuse facilities, which impose higher standards for effluents, electricity contributes about 68%–92% of emissions (footnote 43).

Decarbonizing the water sector therefore requires decarbonizing its electricity supply and replacing inefficient and diesel-powered pumps with high-efficiency electrical pumps. Transportation of sludge must similarly be decarbonized, preferably with electric vehicles powered by renewables.

Transitioning away from fossil-fueled thermal power generation, especially coal-fired power generation, in favor of renewable energy is an essential step. Solar and wind, increasingly supported by energy storage, are critical technologies alongside hydropower and pumped storage. In particular, wind has the lowest emissions of any technology, with onshore wind at 11 grams of CO_2 equivalent per kilowatt-hour and offshore wind at 12 grams of CO_2 equivalent per kilowatt-hour (footnote 32).

Jiamisu cityscape. Municipal District Energy Infrastructure Development Project in the People's Republic of China (photo by Deng Jia).

---

[104] J. Knutsson and P. Knutsson. 2021. Water and energy savings from greywater reuse: a modelling scheme using disaggregated consumption data. *International Journal of Energy and Water Resources*. 5. pp. 13–24.

Policies that support the development of and transition to solar and wind and other low-carbon electricity sources, such as feed-in-tariffs, concessionary interest rates, guaranteed interconnection to the grid, and streamline permitting, will help speed this transition (footnote 100).

4.4 Water Resource Management Policies

Water resources management is a powerful tool for decarbonizing water reservoirs. Government policies are essential to motivate and facilitate efforts to decarbonize water reservoirs, especially because reservoirs present a common problem that individual stakeholders are unable and typically unwilling to bear the cost of solving without assistance and coordination.

Several policies can help promote enhanced reservoir management with the aim of decarbonization. These can be generally categorized as regulatory standards, the provision of tools and technical assistance, financial incentives and/or penalties, and stakeholder coordination.

Regulatory standards to promote enhanced reservoir management must start with the development of information specific to the particular reservoir ecosystem. Requiring reservoir operators to conduct environmental assessments that incorporate the life-cycle analysis across a range of relevant pollutants, including GHG emissions and precursors to GHG emissions (such as nitrates and phosphates), will provide operators, regulators, and other stakeholders the information required to implement sustainable reservoir management.

Regulation should prohibit conversion or degradation of wetlands, which are critical both ecologically and to sequester carbon. Government policy can further require through regulation

that operators minimize impacts identified in the environmental assessment by designing and operating reservoirs using best available technology and best practices.

Decarbonization presents reservoir operators with new and additional challenges that they are likely unfamiliar with and unprepared to act upon. Tools and technical assistance will be necessary to assist operators and other stakeholders in understanding the options available to decarbonize reservoirs. Examples of tools include conducting training for operators and other stakeholders who must gain an understanding of the reservoir GHG emissions pathways and options for decarbonization. The provision of geographic information system tools and services, and artificial intelligence-enabled assessments of optimal locations to site reservoirs in order to reduce emissions can assist planning and government agencies responsible for siting or approving siting of reservoirs.

Policy should incorporate financial support, including incentives for compliance and possibly penalties for noncompliance with government regulation and standards. Incentives for meeting and exceeding standards for decarbonization or taking decarbonization measures should be considered to motivate action. Incentives are especially appropriate for early action on meeting new standards. Financial support should also include support for adoption of advanced technologies, such as reservoir aeration equipment or CH_4 capture mechanisms (footnote 82), and improved practices such as conservation tillage, crop rotation, and cover cropping that can reduce soil erosion and avoid fertilizer runoff into reservoirs (Executive Summary, footnote 8).

Finally, government policy also has the unique ability to coordinate the efforts of stakeholders across a country, region, or reservoir basin. Coordination, aided by the ability to regulate, can help overcome barriers to collective action, which will be essential to implement decarbonization measures in the water resources management subsector.

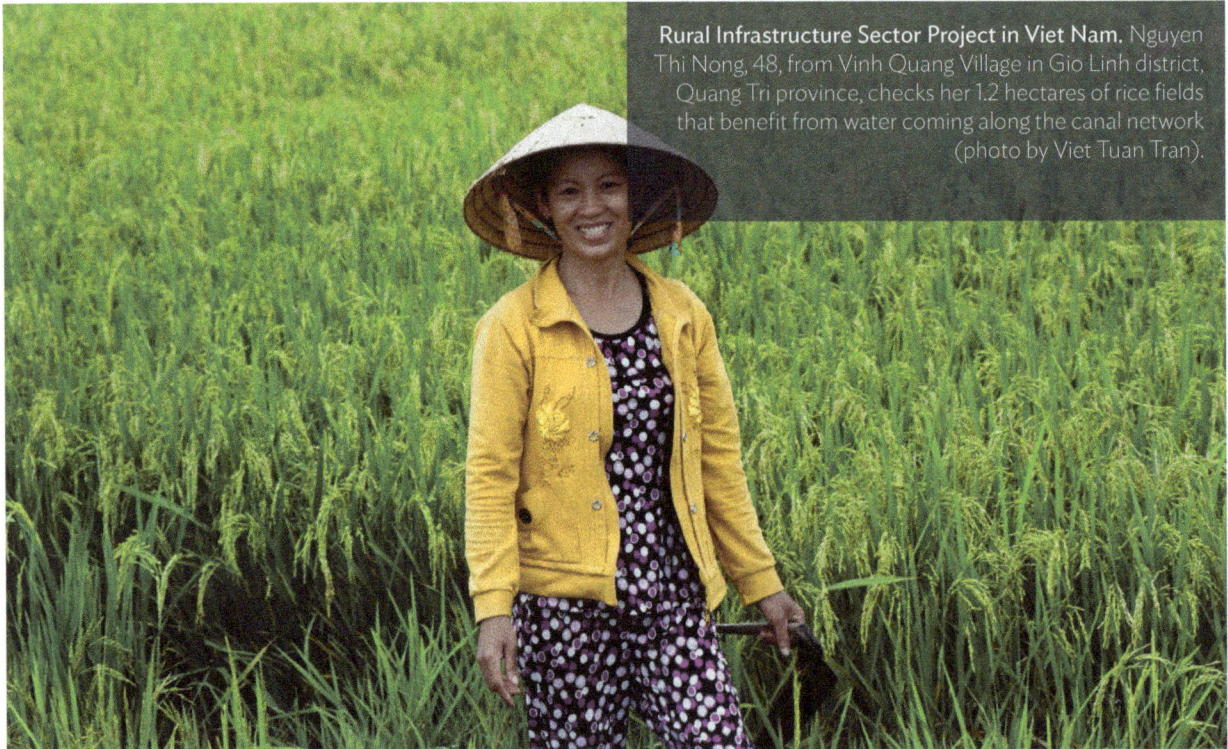

Rural Infrastructure Sector Project in Viet Nam. Nguyen Thi Nong, 48, from Vinh Quang Village in Gio Linh district, Quang Tri province, checks her 1.2 hectares of rice fields that benefit from water coming along the canal network (photo by Viet Tuan Tran).

4.5 Irrigated Agriculture Policies

Agriculture accounts for an estimated 70% of global water consumption (Executive Summary, footnote 6). For this reason, agriculture ranks among the most impactful opportunities to achieve emissions reductions. Several policies can help decarbonize the water sector with measures specific to agricultural consumption.

Agricultural Water Pricing

As described in section 4.2.1, water pricing should encourage the efficient use of water, consistent with national priorities. For centrally provided water, agricultural water pricing should reflect efficiency goals and reward conservation. Subsidized agricultural water consumption promotes waste and encourages agricultural activities that may not be environmentally or financially sustainable. Ending subsidies and internalizing the full costs of providing water on a life-cycle basis are important steps in both efficient use of water and sound economic development.

Crop Substitution and Diversification

Promoting crop substitution should be considered for high-emissions and/or high-water-consumption crops. Replacement crops should be high-value crops that have both market value and meet basic nutritional needs. Other desirable qualities of replacement crops include adaptability to changing climactic and environmental conditions, such as resilience to salt water, drought, pests, and disease.[105] Substitution by individual farmers also helps diversify crops, thereby reducing regional dependency on any particular crop, and thereby enhancing overall food security.

---

[105]   Y. Oladosu et al. 2019. Drought Resistance in Rice from Conventional to Molecular Breeding: A Review. *International Journal of Molecular Sciences.* 20 (14).

Agricultural Water Conservation

Water conservation in the agriculture sector should be actively promoted through training and technology support. Water conservation approaches should be promoted, such as drip irrigation. For example, with the support of ADB, Viet Nam has modernized its irrigation systems in five drought-prone provinces in its south central coastal and central highland regions. These upgrades not only conserve water, but also make farmers and food security more resilient to water shortage.[106]

For rice, the SRI minimizes water consumption by keeping soil moist, instead of flooding crops. This system has been adopted in more than 20 countries throughout Asia and the Pacific (footnote 87). Improvements to agricultural transport and storage infrastructure will help prevent food loss, and thereby reduce water consumption and GHG emissions (footnote 37).

In coping with severe water shortage, India promotes rainwater harvesting, recharging of underground aquifers, construction of check dams, and other water conservation measures (footnote 48), which can serve as a model for other countries in Asia and the Pacific.

These water-saving measures can be combined with other GHG-saving measures like low- or no-till farming, and training to avoid overuse of chemical fertilizers.

Preventing Overfertilization

Overfertilization of crops produces chemical runoff that contaminates water sources, causes eutrophication, and increases the wastewater processing requirements for waters entering these facilities.

Fertilization increases levels of phosphorus and nitrogen in runoff waters. Removal of anthropogenic nitrogen requires additional anaerobic processing that emits N_2O, a potent GHG.

Wetlands protection. Mangroves growing along the beach of Tarawa in Kiribati (photo by Eric Sales).

---

[106]   ADB. 2020. *Irrigation Systems for Climate Change Adaptation in Viet Nam*. Manila.

Preventing overfertilization through training, monitoring, and application technologies, and, if necessary, enforcement actions can help protect water resources and reduce carbon emissions by reducing the need for secondary treatment processes at WWTPs as well as avoiding CH_4 emissions caused by eutrophication of natural water bodies.

4.6 Land Use and Forestry Resource Management Policies

Integrating Water and Wetlands Protection into Land Use Decisions

Decarbonizing the water sector and ensuring the availability of water resources requires a holistic approach that incorporates land use planning, water resource protection, economic development, and urban development into planning water infrastructure and development.

Elemental to a holistic approach, decarbonizing the water sector in Asia and the Pacific specifically requires protecting forests and wetlands, including peatlands, mangroves, and eelgrass, as these are the earth's natural water, sanitation, and wastewater systems that help ensure clean and safe water, and protect ecosystems and water cycles. To the extent that the carbon-free water services they provide avoid the need to build human-made water infrastructure, and the fact that their carbon sink capacity can be preserved and possibly enhanced, the protection of these resources helps decarbonize the water sector.

Land use and development planning should integrate water resources in decision-making and government approval. Adopting integrated water resources management with climate resilience and GHG emissions reduction as objectives helps balance development objectives with the imperative of protecting water resources that provide a service in support of communities.

A comprehensive and rigorous approach to urban development must integrate the entire water cycle, including natural resources, chemicals and energy inputs, and resulting wastewater pollution (footnote 90). Integrated water resource planning should require developers to support their proposals with water source plans, spatial and building designs that conserve water, and water carbon footprints for the proposed development. This analysis will inform government policy and approvals, and public and private investment decisions.

Integrated water resources management embraces life-cycle-assessment-driven analysis, consistent with the monitoring , reporting, and verification approach to measurement of emissions proposed in this section. This approach will enable consideration of various options for achieving water decarbonization and resilience, including decentralized water systems (when they offer favorable energy and performance).

4.7 International Cooperation

Water resource systems often do not respect national boundaries. Transboundary water resources are, by their geography, best managed on a coordinated whole-resource basis, rather than in piecemeal fashion by nations with uncoordinated, or even conflicting, goals and policies.[107]

Lack of coordination over water resources leads to economic loss and even conflict.[108] Institutions such as ASEAN, the Mekong River Commission, and ADB's Central Asia Regional Economic

---

[107]   E. Ostrom. 1990. *Governing the Commons: The Evolution of Institutions for Collective Action.* Cambridge, United Kingdom and New York: Cambridge University Press.

[108]   Adelphi Consult and Central Asia Regional Economic Cooperation. 2017. *Rethinking Water in Central Asia: The costs of inaction and benefits of water cooperation.* Bern: Swiss Agency for Development and Cooperation.

Cooperation Program can play an important role in promoting regional approaches to water resources management, with a view to utilizing these resources to maximize water security, economic development, and decarbonization outcomes.

Additionally, UNFCCC cooperative mechanisms and the Ramsar Convention on Wetlands provide support for international cooperation in decarbonizing and adapting the water sector.

International institutions will play an important role in promoting water decarbonization. This section focuses on two international treaties: the UNFCCC's Paris Agreement and the Ramsar Convention on Wetlands.

United Nations Framework Convention on Climate Change Paris Agreement Article 6 Mechanisms

The Paris Agreement provides a renewed opportunity to elevate the importance of the water sector in mitigating GHG emissions, as well as to synergize mitigation efforts with adaptation measures. The Paris Agreements supports countries in initiating their own mitigation approaches sector-wide through nationally determined contributions (NDCs). The Paris Agreement's Article 6 cooperation mechanisms for mitigation outcomes will help financially support the implementation of NDCs.

NDCs—the foundation of Paris Agreement efforts to decarbonize and adapt to climate change— have to date largely ignored water as a factor in decarbonization planning (footnote 1).

Further, the UNFCCC Kyoto Protocol's CDM, which is the most comprehensive library of applied knowledge concerning decarbonization, and serves as a starting point for the development of Paris Agreement Article 6 methodologies, has developed few methodologies focusing on water

operations. Although CDM methodologies have single-mindedly focused on decarbonization, and the CDM has developed hundreds of methodologies to mitigate GHG emissions and enhance sinks across various infrastructure and sectors, fewer than 20 methodologies specifically decarbonize water sector infrastructure. Water may be considered in these methodologies as a broader indicator of the sustainability of a project, but subordinates water to a supporting factor, without prioritizing water as an opportunity for mitigation outcomes.

To promote water decarbonization with Paris Agreement support, therefore, greater effort must be concentrated on developing the methodologies required to realize Paris Agreement Article 6 support for water infrastructure. For example, the high physical (real) loss for a water utility is often the direct result of a limitation of resources available to implement leakage reduction strategies. The implementation of a carbon balance as part of the water utility's standard water balance assists in identifying and reducing leakage. Documented reduction of physical loss could be used to generate carbon credits, which can then be sold to organizations seeking to achieve carbon neutrality. This new revenue stream could serve to fund further improvements to a utility's infrastructure that may not happen without additional funding. Alternatively, by forecasting an amount of carbon reduction over a number of years, a project can also seek financing for improvements with the promise of reducing carbon emissions over a specific time frame.[109]

Ramsar Convention on Wetlands

The Ramsar Convention on Wetlands is an international treaty that provides a framework for national action and international cooperation for the conservation and wise use of wetlands and their resources.

---

[109] International Water Association. 2023. *Leakage Emissions Initiative: Establishing a Standard Carbon Balance for Drinking Water Utilities.* London. https://iwa-network.org/news/water-loss-specialist-group-white-paper-leakage-emissions-initiative/

Wetlands, such as peatlands, and Arctic permafrost are central to both water and carbon. Peatlands, despite only accounting for a few percent of land, store an estimated one-third of all land-based carbon. Wetlands such as salt marshes, mangroves, and eelgrass beds, also are critical ecosystems that store significant carbon.[110] Arctic permafrost, one of the planet's most important long-term carbon sinks, threatens to release the estimated 1,700 billion metric tons of frozen carbon it stores[111] in the form of CH_4 and CO_2 if climate change continues unabated.[112]

The Ramsar Convention is an important forum for promoting international action and coordination on wetlands, especially shared wetlands among countries. The loss of roughly a third of wetlands in the last 45 years between 1970 to 2015 contributes to global warming by transforming wetland natural carbon sinks into emission sources.

Ramsar Convention stakeholders are actively working to protect wetlands in Asia and the Pacific and encouraging countries to include wetland conservation and restoration in their national policies on climate change, in order to achieve Paris Agreement goals (footnote 110).

[110] Ramsar Convention on Wetlands. 2019. Statement by Martha Rojas Urrego, Secretary General of the Ramsar Convention on Wetlands. "Wetlands: a natural solution to climate change." 30 January. https://ramsar.org/news/wetlands-and-climate-change.
[111] K. R. Miner et al. 2022. Permafrost carbon emissions in a changing Arctic. *Nature Reviews Earth and Environment*. 3. Pp. 55–67.
[112] M. Brouillette. 2021. The Buried Carbon Bomb. *Nature*. 591: pp. 360–362.

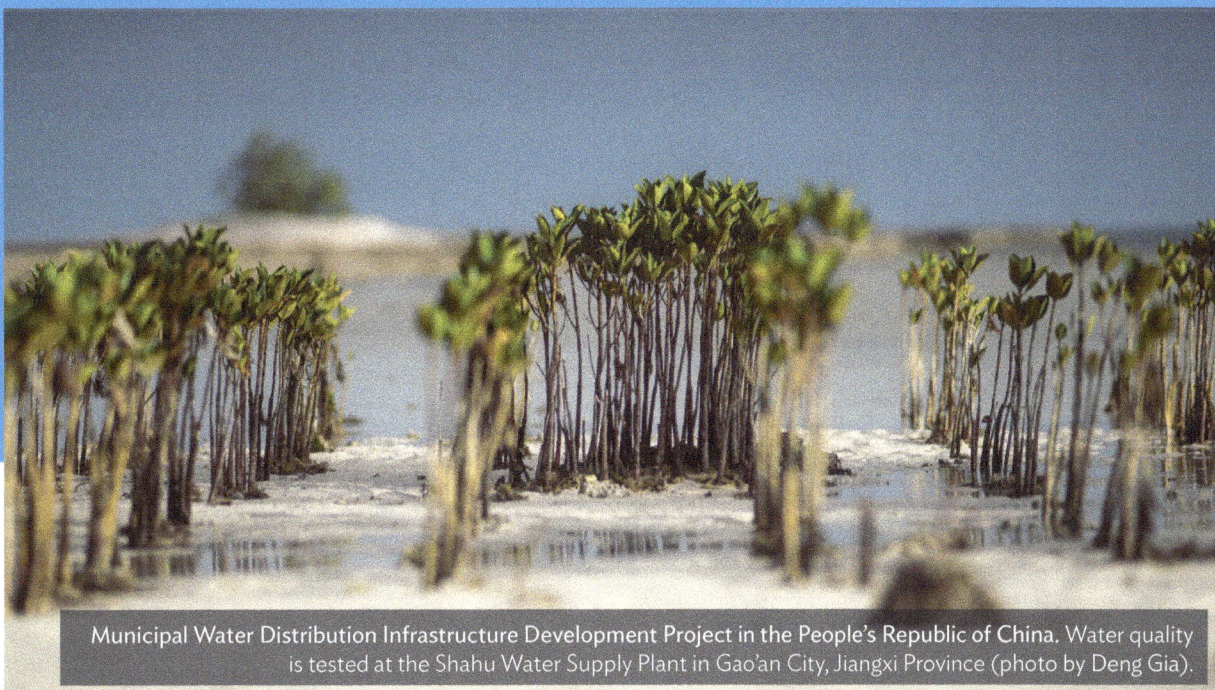

Municipal Water Distribution Infrastructure Development Project in the People's Republic of China. Water quality is tested at the Shahu Water Supply Plant in Gao'an City, Jiangxi Province (photo by Deng Gia).

5 A Framework for Decarbonizing the Water Sector in ADB Operations

A coherent approach to decarbonization in water sector operations is needed to ensure that ADB can effectively support and enable water projects and investments in Asia and the Pacific leading to a net-zero transition, while also achieving resilience outcomes. In this chapter, six mutually reinforcing pillars are proposed as a framework to support this gradual shift, pillars which are complementary to those presented in the guidance note *Mainstreaming Water Resilience in Asia and the Pacific* (Figure 9).

The proposed approach is integrated and inclusive and focuses on long-term climate mitigation outcomes through scaling up financing and building capacity starting with early engagement of DMCs. The approach provides guidance to the water sector in general, and to ADB in developing a portfolio of water projects and programs with a GHG mitigation outcomes, and support DMCs in Asia and the Pacific in their net-zero transition.

The pillars would see ADB

(i) accelerate early engagement and build demand for water sector investments leading to mitigation outcomes;
(ii) adopt a water community approach to DMC decarbonization capacity;
(iii) strengthen ADB staff capacity;
(iv) foster knowledge, innovation, and partnerships, primarily through leveraging the bank's strategic position as a center for thought and practice leadership on water resilience and climate mitigation within Asia and the Pacific;
(v) mobilize finance for water sector decarbonization; and
(vi) spearhead digitalization for decarbonization in the water sector.

Figure 5: Six Pillars Highlighting ADB Actions to Decarbonize the Water Sector

PILLAR 1 Accelerate upstream engagement and build demand for mitigation-focused water sector investments

PILLAR 2 Adopt a water community approach to build decarbonization capacity

PILLAR 3 Strengthen ADB staff capacity

PILLAR 4 Foster knowledge, innovation, and partnerships

PILLAR 5 Mobilize finance for water sector decarbonization

PILLAR 6 Spearhead digitalization for water sector decarbonization

ADB = Asian Development Bank. Source: ADB.

5.1   Pillar 1: Accelerate Upstream Engagement and Build Demand for Mitigation-Focused Investments

To increase the demand for water investments or investments in DMCs that are integrated or cross-sector in nature with GHG mitigation outcomes, decarbonization objectives need to be introduced as early as possible into the project cycle. The indicative country pipeline and monitoring report, listing ADB's project pipeline for the next years, should reflect broad commitment to decarbonization through inclusion of development projects designed using GHG mitigation principles and finding opportunities, where possible, for mutual strengthening projects focused on resilience adaptation. The indicative country pipeline and monitoring report should increasingly reflect projects designed specifically to address both mitigation and resilience needs of communities, regions, and economic sectors in a wide range of potential climate-, economic-, health-, social-, and environment-related shocks.

A logical starting point for building the pipeline of water sector projects and programs with mitigation outcomes is to influence project design as far upstream in the process as possible, beginning with sector assessments and road maps (and their associated documents), which are prepared for sectors, regions, and DMCs, and are designed to inform ADB's country partnership strategy (CPS). Corporate climate financing targets and alignment with the Paris Agreement[113] provide additional impetus for such an approach.

The CPS remains the most important mechanism for introducing climate mitigation in water sector projects as a priority, increasing upstream engagement and developing a robust pipeline of projects with mitigation outcomes. The CPS can be negotiated to emphasize both established approaches toward resilience and decarbonization, including (i) practices to enhance water efficiency and conservation,

---

113   ADB. 2021. ADB Commits to Full Alignment with Paris Agreement. News release. 8 July. https://www.adb.org/news/adb-commits-full-alignment-paris-agreement.

(ii) a clean energy transition, (iii) a reduction in GHG emissions from sanitation and hygiene, (iv) water–food–energy nexus analyses, (v) land use practices to preserve peatlands and mangrove forests, and (vi) nature-based solutions. Once priorities and modalities are defined and agreed, the CPS can then be used as the basis for joint development of a pipeline of projects oriented toward mitigation and resilience.

Actions and approaches. ADB projects are driven by DMC demand. To increase demand, a shared understanding is needed with DMCs on the need for mitigation projects, especially the conditions conducive for finance eligibility. This can be done through workshops and training by ADB, by ensuring this topic is on the agenda at every annual country programming mission, and through a sustained outreach program. These actions can be implemented through the Asia and the Pacific Water Resilience Hub (pillar 4). An example is the Strategy and Partnerships Team's 2023 training course on "Resilience and Decarbonization" for water utilities aiming at building the capacity of water utilities in developing road maps and including short-, medium-, and long-term actions toward enhancing resilience and decarbonization in their operations.[114] Where demand has been generated, support can be enhanced through the provision of technical assistance for project preparation.

At a strategic policy level, coordination between ADB and water issues represented in climate action planning and climate policy documents can be initiated in the near term. NDCs—the foundation of Paris Agreement efforts to decarbonize and adapt to climate change—have, to date, largely ignored water as a factor in decarbonization planning (footnote 1). ADB will work with DMCs to ensure NDCs are updated to include climate mitigation and resilient water management approaches.

5.2 Pillar 2: Adopt a Water Community Approach to Build Decarbonization Capacity

As sovereign borrowers are ultimately responsible for implementing ADB-financed sovereign projects and programs, the awareness and capacity of DMCs are essential preconditions for the successful integration of climate mitigation in the water sector. Even with significant ADB and other expert inputs, successful implementation requires an understanding of national and local contexts to ensure buy-in. Enhanced and expanded capacity in DMC counterparties to conceive, design, and implement mitigation dimensions into water investments is an important step toward the development of an effective ADB water mitigation strategy.

Actions and approaches. Capacity within ADB to conceptualize and implement mitigation water sector policies, projects, and programs can serve as an anchor for an outreach program to build DMC capacity, in partnership with other organizations embodying the required expertise. A quick-start approach for ADB to conduct DMC outreach and capacity building will include the bank doing the following:

(i) Supporting and facilitating activities in DMCs focusing on GHG emission reduction strategies in the water sector through the Asia and the Pacific Water Resilience Hub (pillar 4).
(ii) Setting up a program to reach out to institutions in Asia and the Pacific (and in some cases, globally) that have been innovators in this area to develop working relationships and a coordinated agenda.

---

[114] ADB. Asia and the Pacific Water Resilience Hub. Resilience and Decarbonization: Capacity Building and Roadmap Preparation for Water Utilities (training course). https://hub4r.adb.org/trainings/resilience-and-decarbonization-capacity-building-and-roadmap-preparation-water-utilities.

(iii) Establishing water-centered climate mitigation partnerships with key development and climate finance organizations, bilateral organizations, nongovernment organizations, and knowledge institutes. Many funding agencies may offer joint capacity-building opportunities in climate mitigation in the water sector. Existing relationships with the joint multilateral development banks around climate finance accounting and the Paris Alignment[115] may provide openings for collaboration. Exploring targeted relationships with the United Nations and other organizations already working in this area is another possibility.

(iv) Showcasing mitigation in water sector investments at international forums (such as the annual Conference of Parties and the regional UNFCCC Asia-Pacific Climate Week) and water events (such as the Asia-Pacific Water Forum, Stockholm World Water Week and Singapore International Water Week). Side events and training workshops targeting DMCs in collaboration with partner organizations can be tailored to audiences and focus on research and technical, operational and design, or finance aspects.

(v) Providing resources and support to DMC counterparts and organizations, including the private sector, to rapidly build their capacity for planning water sector projects and programs, with a focus on mitigation and resilience.

5.3 Pillar 3: Strengthen ADB Staff Capacity

The considerable experience and insight into GHG emission reduction in water systems that already exists within ADB can be used as a basis for discussions with DMC counterparts, to build pipelines of water investments with mitigation and resilience outcomes. This experience and insight will evolve as new knowledge is acquired. ADB's in-house capacity can be strengthened by both training and mentoring ADB staff and long-term consultants, and employing skilled specialists.

Actions and approaches. The most rapid and efficient way to expand ADB's internal resources is via direct engagement of experts in the requisite area. An example is the setting up of a panel of experts through the Mainstreaming Water Resilience in Asia and the Pacific TA project[116] and their engagement through resource person contracts. The 30 experts engaged covered disciplines ranging from climate mitigation, adaptation, climate risks, economics, Paris Alignment, strategic water sector analysis, and digitalization. The panel is being extended to also include experts on sanitation and hygiene, and additional expertise may be added in the future depending on demand. The experts provide support to ADB sectors group and private sector operations department as well as technical support for short training and knowledge events both for ADB staff and DMCs.

---

[115] World Bank. The World Bank Group and Paris Alignment. Joint MDB Paris Alignment Approach. https://www.worldbank.org/en/publication/paris-alignment/joint-mdb-paris-alignment-approach.

[116] ADB. Regional: Mainstreaming Water Resilience in Asia and the Pacific. https://www.adb.org/projects/55064-001/main.

Specific suggestions include the following:

(i)   Circulate lessons learned and best practices from water sector projects aiming at reducing GHG emissions and enhancing system resilience, through internal knowledge events and with support of the "Asia and the Pacific Water Resilience Hub" (Pillar 4).

(ii) Preparation of internal tools and guidance documents for ADB practitioners with focus on the quantification of emissions from water sector projects and possible strategies for decarbonization.

5.4 Pillar 4: Foster Knowledge, Innovation, and Partnerships

To some degree, ADB already occupies a unique position in the water resilience space as a facilitator of knowledge. The Asia and the Pacific Water Resilience Hub (hub4r.adb.org), launched in 2022 as a center to share knowledge, tools, and trainings among partners and ADB colleagues, has proved extremely successful. By May 2023, more than 1,200 individuals and 44 organizations had joined the hub and 114 individuals had been trained on topics around water and resilience. However, ADB work around climate mitigation in the water sector is still relatively limited.

Asia Pacific Water Resilience Hub
CONNECT. COLLABORATE. CAPACITATE.

Asia and the Pacific Water Resilience Hub. Managed by the Strategy and Partnerships Team, the hub is an open platform dedicated to strengthening water security in the region.

Actions and approaches. ADB will take advantage of its strategic position on water and resilience to foster knowledge, innovation, and partnerships on climate mitigation in the water sector, possibly in combination with

climate adaptation measures, through the Asia and the Pacific Water Resilience Hub.

Demand-driven training will be provided on climate mitigation in water sector projects as well as on nexus approaches to promote mitigation outcomes (e.g., water–food–energy nexus). Tailored training courses will be developed and provided to selected water partners in DMCs and ADB staff for implementing specific climate mitigation interventions. The courses will be certified by ADB and the partner organization.

To achieve these objectives ADB launched the Asia and the Pacific Water Resilience Initiative at COP27, commonly known as "RUWR: aRe yoU Water Resilient?", a multipronged endeavor to support local-level authorities in its DMCs to mainstream water security and resilience through grant and technical assistance mobilization, and rapid building and sharing of knowledge, tools, and solutions toward resilience.

5.5 Pillar 5: Mobilize Finance for Water Sector Decarbonization

Financing water sector decarbonization is a critical bridge in financial decision-making to reallocate capital from carbon-intensive, fixed climate water-intensive assets, products, and services. This can move the region toward a more resilient, low-carbon future and help it endure the effects of climate change. Financing can be achieved through a wide range of options, including concessional finance; private sector investment (businesses, pension funds, commercial banks, investor groups); grants; alternative financing instruments (such as bonds and blended finance); and revenue-generating services. In addition, different financing modalities (e.g., policy-based loans, sector development programs, results-based loans, and project readiness financing) can provide opportunities to stimulate reforms, providing incentives to strengthen mitigation and adaptation.

BOX 5

Sri Lanka as an Example

Sri Lanka is discussing with the Asian Development Bank a water sector policy-based loan to implement a series of reforms to address vulnerabilities for improved sector performance, efficiency, resilience, and environmental sustainability. One of the policy actions under the proposed program consists of preparing a climate change and resilience road map for the main water service provider in the country to mainstream climate resilience through short-, medium-, and long-term targets on climate adaptation, energy efficiency and/or saving, renewable energy, greenhouse gas reduction, carbon footprint reduction, water loss reduction, and avoided methane emissions. This would include an indicative financing plan for the identified and mitigation actions, supplemented with institutional actions (soft components) needed to enhance climate change awareness and resilience. Another policy action comprises developing a framework, strategy, and road map for increasing private sector participation in the service provider, including an action plan to enhance environmental, social, and governance sustainability into the service provider's overall plans and operations, therefore building its institutional capacity to apply for eligible green climate financing.

Source: Asian Development Bank.

Actions and approaches. ADB will encourage creative approaches to aligning climate mitigation in water sector projects with the investment and partnership process. ADB will continue to work with DMCs (nationally and locally); other multilateral organizations; bilateral aid groups; and civil society organizations to unlock finance from ADB-managed funds toward mitigation and resilience (e.g., the Water Financing Partnership Facility, Urban Climate Change Resilience Trust Fund, and the Japan Fund for Prosperous and Resilient Asia and the Pacific). ADB will also work with external funds such as the Green Climate Fund and the Bill & Melinda Gates Foundation.

The private sector can play an important role in providing the required resources, financing, and know-how to foster the decarbonization of the water sector. As discussed in Chapter 4, examples of activities where the private sector could play a role may include the following, among others: (i) deploying affordable, high-efficiency technologies such as "smart" pumps,[117] leak detection sensors, and other digitally powered solutions that dramatically reduce the amount of energy used in the treatment and transport

of water; (ii) deploying technologies for optimization of WWTPs and including carbon capture and energy recovery; (iii) deploying measuring technologies to quantify baseline emissions; and (iv) purchasing carbon credits from water sector operations to offset other carbon-intensive operations.

5.6 Pillar 6: Spearhead Digitalization for Water Sector Decarbonization

The adoption of technology has been frustratingly slow in the water sector, despite evidence of the success of digital solutions across a wide range of applications. Information from ADB's Digital Technology Unit on digitalization projects or project components approved from 2010 to 2020 shows the water sector ranked one of the lowest in digitalization uptake.

Actions and approaches. To spearhead digitalization in water sector projects in DMCs, ADB will foster collaboration with public and private development partners and solution

---

[117] "Smart" pumps are pumps that have the ability to regulate and control flow or pressure. Typical advantages are energy savings, lifetime improvements, and system cost reductions.

providers. In March and October 2021 and June 2023, ADB organized its first three e-marketplaces for water management, bringing together providers of information and communication technology, digital and remote sensing tools, and technologies applicable to all subsectors of water management in its DMCs. ADB plans to continue such e-marketplaces to connect ADB's water sector staff and DMC clients and stakeholders with innovative technological solutions from the market. ADB will also support production of knowledge products that promote digital solution providers and tools and technologies on water sector decarbonization and resilience.

ADB will support digitalization initiatives and help its DMCs create an enabling environment, including via human resources and capacity, to upscale and mainstream digitalization for promoting mitigation in the water sector. ADB will support integration of information and communication technology, digital, and remote sensing technologies in three ways:

(i)   **Financing and investment.** ADB provides financial assistance to countries to incorporate digitalization in water- related projects.

(ii) **Technical assistance.** ADB offers technical expertise and advisory services to its DMCs to strengthen their capacity in implementing digitalization in water-related projects. This includes conducting studies, providing training and knowledge sharing, and supporting the development of policies and regulations.

(iii) **Partnerships.** ADB collaborates with various stakeholders, including governments, nongovernmental organizations, and other development agencies, to promote digitalization for water. These partnerships help in sharing best practices, fostering regional cooperation, and mobilizing resources.

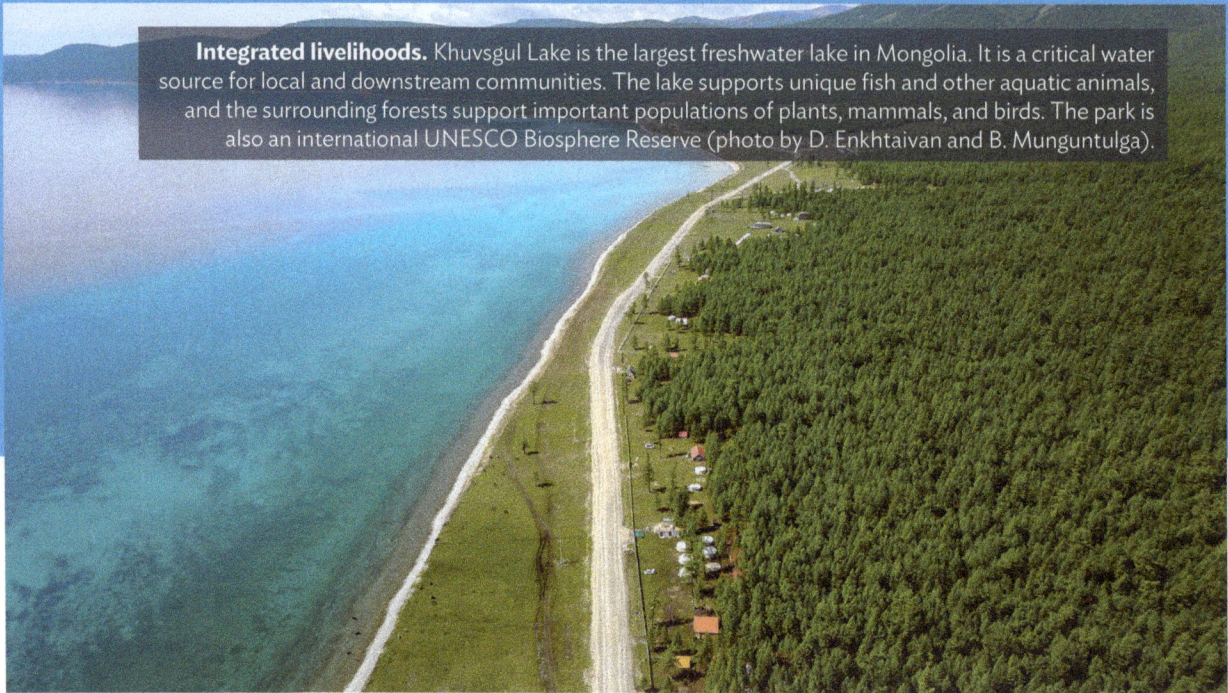

Integrated livelihoods. Khuvsgul Lake is the largest freshwater lake in Mongolia. It is a critical water source for local and downstream communities. The lake supports unique fish and other aquatic animals, and the surrounding forests support important populations of plants, mammals, and birds. The park is also an international UNESCO Biosphere Reserve (photo by D. Enkhtaivan and B. Munguntulga).

6 Summary

The water sector has been widely overlooked as a significant source of GHG emissions and consequently has often been excluded from climate mitigation planning and reporting. This is mainly the result of a lack of knowledge on and the difficulty in accurately quantifying GHG emissions across all subsectors of the water sector. As a result, the water sector has also been overlooked as an opportunity to mitigate climate change.

This guidance note discusses technical, institutional, financial, and policy-related opportunities that may foster decarbonization in the water sector, including examples and good practices from the Asia and Pacific region that could be replicated and scaled up to support the net-zero transition. In assessing opportunities to mitigate GHG emissions while enhancing the resilience of the water sector, this guidance note evaluates the five water-related subsectors in which ADB is active: water supply; wastewater, sanitation, and hygiene; water resources management, energy, and reservoirs; irrigated agriculture; and land use and forestry resource management. The document also provides concrete actions under six mutually reinforcing pillars that can set the basis for ADB to further enhance mitigation actions in the water sector through collective actions in cooperation with partners and DMCs.

Proper water management, planning, and action can substantially reduce GHG emissions. Further, opportunities to reduce GHG emissions in the water sector must be included in climate mitigation plans and policies. Priority should be given to measures that conserve water resources, utilize energy efficiently, and protect wetlands (which sequester carbon and enhance biodiversity). It is important that the industry mobilizes finance to fund these efforts, and to strengthen governance to ensure water is mainstreamed into climate mitigation planning and action in the water sector to accelerate the net-zero transition. International cooperation and treaties are essential to drive this transition: mechanisms such as Article 6 of the UNFCCC Paris Agreement and the Ramsar Convention on Wetlands offer opportunities to pursue cooperation to allow for higher ambition in both mitigation and adaptation actions, promoting sustainable development and environmental integrity.

APPENDIX

Technical Overview of Municipal Wastewater Treatment Plant Operations

This appendix describes a typical municipal wastewater treatment plant (WWTP), its basic operations, and sources of greenhouse gas (GHG) emissions. The treatment plant and processes presented are representative of the most common approach to wastewater treatment infrastructure globally.

Figure A.1 shows the operations of a typical municipal WWTP, common throughout the world. The plant's operations involve two distinct processes: treatment of wastewater and treatment of sludge.

Wastewater Treatment Process

Wastewater enters the facility at the headworks. Wastewater first passes through a screen to remove any large foreign materials, such as trash and tree branches. At this stage, the wastewater also passes through a chamber or tank in which the influent is slowed down so that heavy sand and grit settle out.

Next, the wastewater undergoes primary treatment. Primary treatment involves initial clarification through settling of suspended solids, grease, and scum from wastewater, often using chemical flocculant and mechanical separation. A flocculant is a chemical or other substance that, when added to wastewater, traps or attracts particulate suspended solids into clusters or clumps. With the aid of flocculating agents,

Figure A.1: Typical Central Wastewater Treatment Plant

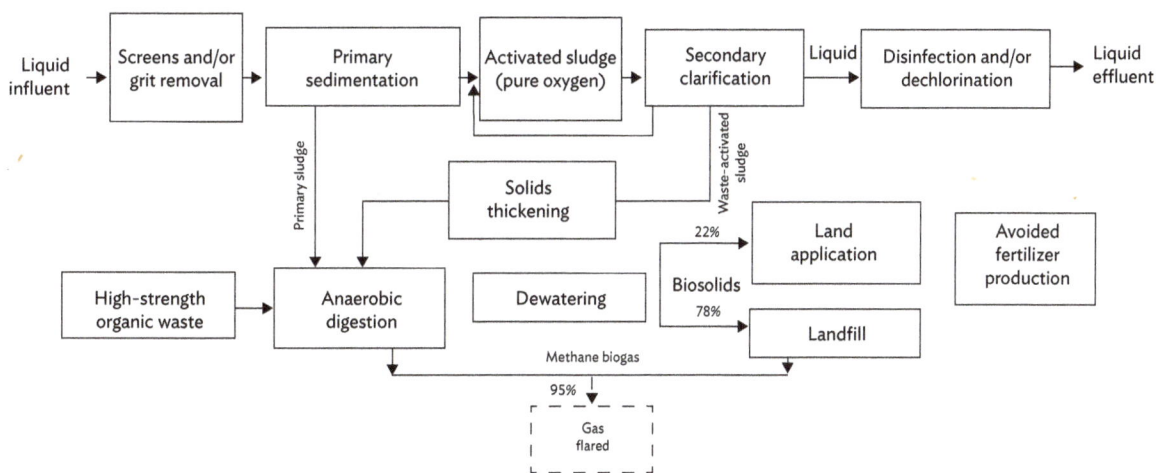

Source: Adapted from A. Shehabi, J. R. Stokes, and A. Horvath. 2012. Energy and air emission implications of a decentralized wastewater system. *Environmental Research Letters.* 7 (2).

primary treatment can eliminate 50%–65% of suspended solids. Solids removed by primary treatment may comprise as much as 30%–40% of the original biochemical oxygen demand (BOD) of the wastewater.[118]

Next, wastewater typically undergoes secondary aerobic digestion treatment employing an activated sludge process. Activated sludge contains microbes that digest organic materials. Sludge removed during treatment is waste activated sludge. The rest of the activated sludge, called return activated sludge, is returned to the aeration tank to maintain the balance between microbes and organic materials.

Aerobic digestion of waste is the natural biological degradation and purification process in which bacteria that thrives in oxygen-rich environments breaks down and digests organic waste present in the wastewater. In the process, large quantities of air are bubbled through wastewater in open aeration tanks to accelerate the digestion process. During the oxidation process, pollutants are broken down into carbon dioxide (CO_2), water, nitrates, sulfates, and biomass (microorganisms). This process is the first step in the nitrification-denitrification process. After several hours in a large holding tank, the clarified wastewater is separated from the sludge of bacteria and discharged from the system. Most of the activated sludge is returned to the treatment process, while the remainder is disposed of by one of several acceptable methods.

To optimize performance of an activated sludge system, the amounts of organic waste (nutrients), organisms (activated sludge), and oxygen (dissolved oxygen) must be balanced. The flow rate of wastewater, the amount of air pumped through the wastewater, the temperature, and the amounts of total suspended solids containing the microbes are coordinated to ensure the optimal

and energy-efficient operation of the primary treatment process. BOD testing is performed periodically to confirm that optimal operating conditions are maintained.

Following secondary treatment, the clarified effluent may require additional aeration and/or other chemical treatment to disinfect and destroy any bacteria remaining from the secondary treatment stage, as well as to increase the content of dissolved oxygen needed for oxidation of residual BOD. Tertiary treatment is often performed using filtration to remove microscopic particles, such as the use of anthracite coal as a filter medium, chlorination, chlorine dioxide, ozonation, or ultraviolet treatment.

Following tertiary treatment, treated water is then released into the environment or reused for agricultural or other consumption.

Nitrification-Denitrification

Nitrification-denitrification may be part of the secondary treatment step in order to remove any organic nitrogen fertilizers in the form of ammonia, which can cause eutrophication of water bodies and can be toxic to fish. Just as BOD indirectly measures the concentration of carbon-based organics in wastewater by measuring the oxygen demand required to produce carbonaceous oxidation of all organics, nitrification is measured by total Kjeldahl nitrogen.

Nitrification-denitrification is commonly performed in separate open tanks or a single open tank with separate zones. Separation or zoning is necessary because different microbes are involved at each step of the process, each having different requirements. The first tank or zone is an aerated basin. The second tank or zone operates as an anoxic zone.

---

[118]   BOD is the quantitative measure of the oxygen needed by bacteria and microorganisms for the biological oxidation of organic wastes in a unit volume of wastewater. BOD is generally measured in milligrams per liter of oxygen consumed over a 5-day period. Ecologix Environmental Systems. Wastewater Terms and Glossary. https://www.ecologixsystems.com/resources-glossary/.

During the first stage, the nitrification stage, a microbial bacterial process is induced by passing air through the wastewater, which causes ammonia nitrogen compounds in the water to combine with oxygen to produce nitrates. Specifically, the aerobic process converts ammonia, ammonium, and similar nitrogen compounds present into nitrite and then nitrate.

In the second stage, the denitrification stage, the wastewater is transferred to the anoxic tank or zone, in which a different set of carbonaceous and nitrogenous microorganisms present at this stage requiring little or no oxygen use chemically combined oxygen in nitrite and nitrate to convert these principally to dinitrogen, nitrous oxide gases, and CO_2.

Anaerobic Decomposition and Sludge Treatment Process

The sludge treatment process can be conceptualized as a distinct process that links to the water treatment process. The sludge treatment process is an anaerobic process.

As organic solids are separated during primary treatment and secondary treatment, mixed effluent and sludge are delivered to an anaerobic digestion tank.

Anaerobic digestion is a process through which bacteria breaks down organic matter present in the effluent and sludge (such as animal manure, wastewater biosolids, and food wastes) in the absence of oxygen in a sealed reactor vessel.

Anaerobic decomposition occurs under strict anaerobic conditions—no oxygen must be present and specific microbes are present that thrive in an anaerobic environment. The anaerobic digestion process breaks down complex organic matter (carbohydrates, proteins, and fats) into basic soluble organic molecules (sugars, amino acids, and fatty acids). These organic molecules then ferment in the anaerobic digestion vessel, producing acetic acid, hydrogen, and CO_2, which further interact to produce primarily methane (CH_4) and CO_2. CH_4 production varies with the volume of organic waste fed to the digester and the digester temperature.

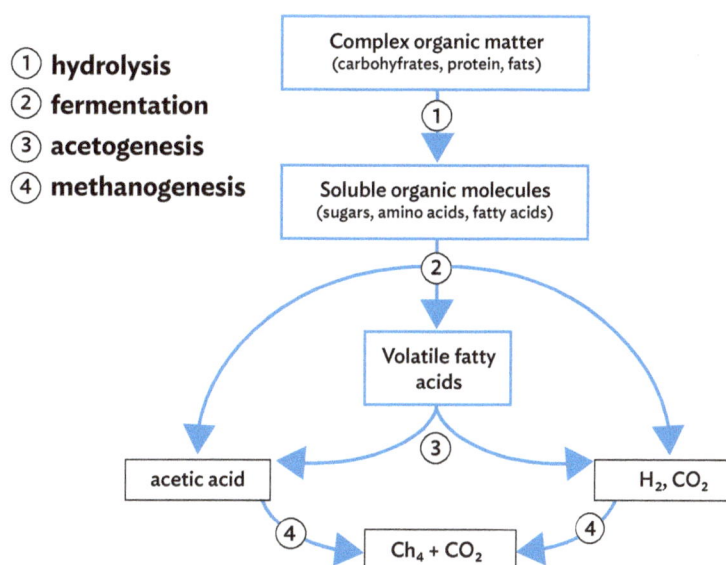

Figure A.2: Anaerobic Digestion Process

① **hydrolysis**
② **fermentation**
③ **acetogenesis**
④ **methanogenesis**

Complex organic matter
(carbohyfrates, protein, fats)

Soluble organic molecules
(sugars, amino acids, fatty acids)

Volatile fatty acids

acetic acid

H_2, CO_2

$Ch_4 + CO_2$

CH_4 = methane, CO_2 = carbon dioxide, H = hydrogen.
Source: A. Costa et al. 2015. Anaerobic Digestion and its Applications. Washington, DC: United States Environmental Protection Agency.

In summary, anaerobic digestion occurs in four steps:

(i) **Hydrolysis.** Complex organic polymer chains are decomposed into simple soluble organic molecules using water to split the chemical bonds between the compounds.

(ii) **Acidogenesis (fermentation).** Chemical decomposition of simple monomers and oligomers of carbohydrates, amino acids, and fatty acids by enzymes, bacteria, yeasts, or molds in the absence of oxygen, producing volatile fatty acids.

(iii) **Acetogenesis.** Acetogenic bacteria convert fermentation products into primarily acetic acid, hydrogen, and CO_2.

(iv) **Methanogenesis.** Acetic acid interacts with hydrogen and CO_2, converting to methanogenic bacteria, producing CH_4 and CO_2.[119]

During this process, CO_2 and CH_4 are released, with CH_4 being vented to the atmosphere, flared, or captured at the top of the reactor and reused to produce heat and/or power. If vented, fossil-derived CH_4 has a global warming potential of 29.8 times that of CO_2 over a 100-year period and 82.5 times over a 20-year period (footnote 54).

Following digestion, waste sludge is dewatered using gravity sedimentation in a lagoon; filter press; chemical additives; or heating and/or evaporation; a mechanical press, centrifuge, or freezing to concentrate the solids. The resulting dry sludge may then be disposed of by landfill or reused as a fertilizer by spreading on farmland.

The anaerobic process offers opportunities for WWTPs to expand services to carbon management by processing food waste, manure, and other organic waste into sludge to reduce CH_4 released in uncontrolled environments and for reuse as fertilizers and building materials, thereby displacing emissions from production of these products.

Phosphorus Removal

Phosphorus, also a fertilizer that causes eutrophication in water bodies, can be separately removed by using biological means during the secondary treatment activated sludge process, or by the introduction of chemicals or coagulants to precipitate the phosphorus as a solid, or a combination of both in the secondary and tertiary treatment stages. For biological removal, organic material must be present to support phosphorus accumulating organisms, which store excess polyphosphate in their cell mass, thereby removing phosphorus from waste sludge. Enhanced biological phosphorus removal uses alternating anaerobic and aerobic zones to accelerate the process. The aerobic process generates water and CO_2 as by-products.

---

[119]   A. Costa et al. 2015. *Anaerobic Digestion and its Applications.* Washington, DC: United States Environmental Protection Agency.